THE FRUITS OF REVOLUTION

THE POLITICAL ECONOMY OF INSTITUTIONS AND
DECISIONS

Editors
James E. Alt, Harvard University
Douglass C. North, Washington University in St. Louis

Other books in the series
James E. Alt and Kenneth Shepsle, eds., *Perspectives on Positive Political
Economy*
Yoram Barzel, *Economic Analysis of Property Rights*
Robert Bates, *Beyond the Miracle of the Market: The Political Economy
of Agrarian Development in Kenya*
Gary W. Cox, *The Efficient Secret: The Cabinet and the Development of
Political Parties in Victorian England*
Leif Lewin, *Ideology and Strategy: A Century of Swedish Politics*
(English Edition)
Gary Libecap, *Contracting for Property Rights*
Matthew D. McCubbins and Terry Sullivan, eds., *Congress: Structure
and Policy*
Douglass C. North, *Institutions, Institutional Change, and Economic
Performance*
Elinor Ostrom, *Governing the Commons: The Evolution of
Institutions for Collective Action*
Charles Stewart III, *Budget Reform Politics: The Design of the
Appropriations Process in the House of Representatives, 1865–1921*

In *The Fruits of Revolution* Jean-Laurent Rosenthal investigates two central questions in French economic history: To what extent did institutions hold back agricultural development under the Old Regime, and did reforms carried out during the French Revolution significantly improve the structure of property rights in agriculture? Both questions have been the subject of much debate. Historians have touched on them in a number of local studies, yet usually they have been more concerned with community conflict than with economic development. Economists generally have researched the performance of the French economy without paying much attention to the impact of institutions on specific areas of the economy. This book attempts to utilize the best of both approaches: It focuses on broad questions of economic change, yet it is based on detailed archival investigations of the impact of property rights on water control.

Part I provides both an introduction to French economic history between 1700 and 1860 and an introduction to the economic literature on property rights and institutions. Part II first looks at water control from a national perspective and then examines two case studies, one of drainage in Normandy and one of irrigation in Provence. The national evidence shows that most water control efforts failed before 1789, whereas 1820–60 were boom years for irrigation and drainage. Quantitative and qualitative evidence suggests that neither technology nor relative prices were responsible for the failure to develop agriculture under the Old Regime; rather, ambiguous property rights, divided authority, and endless litigation all conspired to reduce the efficacy of water control. The Revolution solved important institutional problems in the countryside by centralizing authority over eminent domain, reforming the judiciary, and clarifying property rights to land water. As a result, after 1820 water control flourished.

Part III of the book is devoted to explaining why inefficient property rights arose in the Middle Ages and prevailed until 1789. A set of theoretical models is analyzed to argue that ill-defined property rights were part of the very structure of the Old Regime – a fact that made reform impossible without a revolution.

THE FRUITS OF
REVOLUTION

*Property rights, litigation, and
French agriculture,
1700–1860*

JEAN-LAURENT ROSENTHAL
University of California, Los Angeles

The right of the
University of Cambridge
to print and sell
all manner of books
was granted by
Henry VIII in 1534.
The University has printed
and published continuously
since 1584.

CAMBRIDGE UNIVERSITY PRESS

Cambridge

New York   Port Chester   Melbourne   Sydney

CAMBRIDGE UNIVERSITY PRESS
Cambridge, New York, Melbourne, Madrid, Cape Town, Singapore, São Paulo, Delhi

Cambridge University Press
The Edinburgh Building, Cambridge CB2 8RU, UK

Published in the United States of America by Cambridge University Press, New York

www.cambridge.org
Information on this title: www.cambridge.org/9780521103121

First published 1992
This digitally printed version 2009

A catalogue record for this publication is available from the British Library

Library of Congress Cataloguing in Publication data
Rosenthal, Jean-Laurent.
The fruits of revolution : property rights, litigation, and French
agriculture, 1700–1860 / Jean-Laurent Rosenthal.
p. cm. – (The political economy of institutions and
decisions)
Based on the author's thesis (Ph.D.) – California Institute of
Technology, 1988.
Includes bibliographical references and index.
ISBN 0-521-39220-9

1. France – Economic conditions – 18th century. 2. France – Economic
conditions – 19th century. 3. Right of property – France – History.
4. Drainage – France – Normandy – History. 5. Irrigation – France –
Provence – History. 6. Agriculture and state – France – History.
I. Title. II. Series.
HC275.R59   1992
338.1′0944 – dc20   91-20091
                                                                  CIP

ISBN 978-0-521-39220-4 hardback
ISBN 978-0-521-10312-1 paperback

Contents

Contents

Tables, figures, and maps

Tables, figures, and maps

Series editors' preface

This series, The Political Economy of Institutions and Decisions, is built around attempts to answer two central questions: How do institutions evolve in response to individual incentives, strategies, and choices, and how do institutions affect the performance of political and economic systems? The scope of the series is comparative and historical rather than international or specifically North American, and the focus is positive rather than normative.

This books confronts a historic debate over the influence of the French Revolution on economic change in France. For some scholars, the French Revolution was a fundamental revolution that, by altering the political and economic order, basically changed the direction of French society and economy. For others, the Revolution was simply a violent interruption that, in the long run, had little impact on the direction of French society. In order to probe this issue, Jean-Laurent Rosenthal assesses the differential performance of French agricultural sectors before and after the Revolution. Specifically, he examines irrigation and drainage changes to document both the cost of making such developments in the Old Regime and the cost of making them after the Revolution. The result not only provides detailed substance to the debate about whether the French Revolution was indeed a fundamental revolution but also establishes a new empirical underpinning for the whole property rights literature. This study provides a solid basis for evaluating the way alterations in property rights influence the performance of economies.

Preface

Eighteenth-century French agriculture is often portrayed by historians as shackled by institutional and technological constraints. The countryside was littered with the remains of the feudal regime, and these institutional encumbrances stood in the way of modern market-oriented production techniques. In 1789 the high winds of revolution swept through rural France and shook it violently. For some scholars the tempest of 1789 blew the remains of the feudal regime into oblivion and created a new economic and social order. Others have argued that by 1820 few traces of the storm remained: The French rural world was still hopelessly backward compared with its modern British counterpart. The turmoil of the Revolution had been for naught; only the slow process of urbanization and technological change could relieve the problems of the French countryside.

This book investigates specific sectors of agriculture in France – irrigation and drainage – to document both the costs of Old Regime institutions and the consequences of reforms carried out during the Revolution. In an effort to clarify the connection between institutional reform and economic change, the study uses the methodologies of social science both in its treatment of quantitative evidence and in its explicit use of formal economic models.

In order to make both historians and economists comfortable, I begin in Part I not only by summarizing the relevant historical debates, but also by presenting the economic models that will be used in the analysis. Part II of the book comprises four chapters. First, I provide evidence about the extent of drainage and irrigation efforts between 1700 and 1860 at the national level. Second, I briefly examine relative price changes between 1600 and 1860 to argue that changes in the profits from drainage cannot explain why water control projects failed before 1789 but succeeded after 1820. Next, I turn to case studies of drainage projects in Normandy and irrigation efforts in Provence to show that the most likely explanation of the agricultural failures of the Old Regime lies neither with technology nor with relative prices. Rather, institutions were to blame for the lack of development before 1789. The expansion of drainage and irrigation after 1815 required the legal reforms of the Revolutionary period. Part III is perhaps the most ambitious. There I seek to explain why specific property rights proved so difficult to change under the Old Regime and why they blocked all water control efforts. Instead of relying on archival evidence as in the earlier parts of the book, I base my arguments largely on theoretical analyses of property rights.

This book is the product of a two-step research process. It is a thoroughly rewritten and expanded version of my Ph.D. dissertation, which was completed at the California Institute of Technology in 1988. The dissertation was warmly received and I was encouraged to prepare a manuscript for publication. Yet I was also challenged to address issues beyond the scope of the dissertation. Because I have tried to respond to these challenges, the book is significantly longer than the original dissertation.

While completing this research often involved long periods of solitary toil, it was also marked by a great deal of intellectual generosity and collegiality both from the faculty at Caltech and from my colleagues at the University of California, Los Angeles. The larger community of economic historians has proved to be a place where learning continues long after the dissertation is complete. Therefore, I am indebted to many individuals. First and foremost my gratitude goes to Philip T. Hoffman, my dissertation adviser. Professor Hoffman had the patience and generosity to supervise my research for the better part of three years and initiated me into the convoluted world of French archives. As an economic historian, my ambition is to bridge the gap between economics and history. Professor Hoffman set high standards for me by demanding both constant attention to detail and research focused on big historical questions such as the consequences of the French Revolution. Without him this book would have fallen short as history.

My admiration and thanks also go to Lance E. Davis and Louis L. Wilde. Professor Davis taught me the importance and limitations of quantitative evidence in economic history. He forced me to challenge my own convictions about the importance of institutions; answering his questions strengthened this study and gave me a better understanding of its weaknesses. Professor Wilde initiated me into the modeling of institutional questions as game-theoretic problems. During my foray into law and economics, he kept me focused on the questions I wanted to answer instead of on the mathematics.

Among the historians at Caltech I am most indebted to Eleanor Searle, Morgan Kousser, and the late John Benton. From Professor Benton and

Professor Searle I learned how Old Regime agricultural institutions and property rights had come to be defined between 1100 and 1400. In addition, Professor Kim Border, Professor Jean-Jacques Laffont, Professor John Ledyard, and Professor Thomas Palfrey all offered valuable suggestions as I was struggling with my model of litigation and settlement. I learned game theory from them. I also learned much from Kemal Guler, Shawn Kantor, Mark Olson, and David Porter, my peers in graduate school. My thanks go to all these people.

Since coming to UCLA I have benefited from the advice of Kenneth Sokoloff and Michael Waldman, both of whom generously read the entire manuscript in its penultimate form and offered many helpful suggestions. Finally, the contents of this book have been presented at seminars at Berkeley, Caltech, Chicago, Harvard, Illinois, Massachusetts Institute of Technology, Michigan, Stanford, UCLA, the University of British Columbia, and Washington University. I have thus benefited from the comments of economists and historians too numerous to mention. I especially thank for their interest and criticism those economic historians whose work centers on France – among them George Grantham and David Weir, who read the dissertation and offered extensive comments. Their suggestions, as well as those of Donald McCloskey, greatly improved the book. I thank Douglass North for making institutions such an important part of economic history. Professor North also provided needed editorial guidance and inspiration while I was revising the manuscript.

In France, a number of friends and relatives offered greatly needed lodging and support during the archival campaigns. I owe much to Michel Duplessis-Kergomard, Dominique Kugler, Jöelle Richard, Martine Rubio, and Eric Tamisier. The staff of the departmental archives of the Vaucluse in Avignon was unusually helpful, but I especially thank Jean Mazet, without whose help I could not have collected a data set on wages. Closer to home, my sister, brother, mother, and father, in different ways, helped me weather this long journey. One person, however, had almost constant input into the book: my wife, Paula Scott. Paula not only gave me support throughout this endeavor, but offered editorial advice and supervision. Her constant attention has greatly improved this book. For all these reasons, my greatest thanks go to her.

I would like to thank the division of the Humanities and Social Sciences of the California Institute of Technology, which provided significant financial support over three summers for archival research in France. I accomplished much of the research while I was supported by the John Randolph Haynes and Dora Haynes Fellowship during the spring and summer of 1987 and a Sloan Dissertation Fellowship from October 1987 through June 1988. I was able to collect additional data using funds from the Milton and Francis Clauser Dissertation Prize and the Alexander Gerschenkron Dissertation Prize. At UCLA the International Studies and

Preface

Overseas Programs in conjunction with the Department of Economics funded archival research in 1989 and 1990. This study could not have been completed without such generous financial support.

I dedicate this book to my grandparents, Elinor, Arnold, and Henri, for their friendship and love.

1

Introduction

The Revolution of 1789 was once seen as a watershed; it had, some argued, abruptly transformed France from a backward feudal society into a modern, progressive bourgeois and capitalistic system. This perspective was based on the premise that the Revolution brought about dramatic social and economic change. Yet what was once seen as the birth of modern France is now sometimes described as a bloody struggle for power. The Revolution's social transformation is considered to have been mere redistribution within a now more widely defined elite.

The Revolution coincided with, and was fueled by, the first major harvest crisis in nearly forty years. Riots broke out, possibly because the government was expected to provide relief from the crisis when in fact it lacked both the will and the resources to act. More generally, it was long argued that the Revolution occurred at the end of forty years of failed reform – in areas as wide ranging as agriculture, trade, the judiciary, and taxation. The Old Regime appeared weighed down by inefficient institutions that the state could not alter. In contrast, some recent scholarship has shone a more favorable light on the economic performance of absolutist France. The strongest debate, perhaps, rages over two questions relating to agricultural policy and performance. First, to what extent did institutions curtail increases in agricultural output and productivity in the eighteenth century? Second, to what extent did Revolutionary reforms contribute to the agricultural development of the nineteenth century?

This book examines the importance of the economic reforms of the Revolution by analyzing the role of institutions in determining investment in water control – irrigation and drainage of marshes – both before and after the Revolution. Water control never represented the bulk of investment in agriculture, let alone in the French economy. Nonetheless, its fate is indicative of the problems faced by investors more generally.

1

Introduction

In fact, water control is the kind of investment that magnifies the role of institutions, because providing improvements like drainage and irrigation requires that a score of contracting problems be overcome. Drainage is a local public good in that it is extremely costly to drain only part of a marsh. Thus, some mechanism must exist to ensure that all owners of the marsh participate in the draining effort. Otherwise some owners will be tempted to shirk and force the costs of drainage on the others. Irrigation and drainage also feature externalities – that is, they affect individuals other than the direct participants in the projects. For example, drainage alters the flow of water, which can reduce the value of downstream mill sites. Finally, irrigation is dependent on the provision of rights of way. Thus, water control is acutely dependent on institutions that structure agreements among landowners, on institutions that allocate compensation when individuals suffer damages from a project, and on institutions that decide on eminent domain rights.

By examining the specific institutions related to water control we can show that it was very difficult to answer questions regarding economic change under the Old Regime. If the provision of water control presents an interesting economic contracting problem, it is also an important historical issue. Both irrigation and drainage were means of increasing France's cultivated area, a perennial concern of Old Regime administrators. Royal administrators' attempts to increase investment in irrigation and drainage were very controversial. Thus, we must first judge whether institutions hindered change before 1789. Second, since Revolutionary governments, like Old Regime administrators, focused on institutions as the cause of economic backwardness, we should evaluate their ability to carry out property rights reform. Finally, by following irrigation and drainage activities through the Revolution, we can ascertain whether such costs declined and whether the provision of water control increased as a result of the general institutional change that occurred after 1789.

The study of irrigation and drainage will illuminate the larger impact of the Revolution's reforms. Indeed, both before and after 1789, irrigation and drainage never received much direct attention. As a result, water control benefited only from the general reform process that altered the regime of common land, renovated civil litigation, and curtailed the power of local organizations. Thus, other sectors of the economy must have also felt the effects of the institutional change that benefited irrigation and drainage. For instance, reform of civil litigation lowered the cost not only of drainage, but of contracting in general. Similarly, when localism was reduced during the Revolution, the benefits were felt throughout the economy rather than simply in irrigation.

2

KEY

1 Brittany	12 Orléanais	23 Aunis
2 Normandy	13 Nivernais	24 Gascony and Guyenne
3 Picardy	14 Burgundy	25 Limousin
4 Artois	15 Lorraine	26 Auvergne
5 Flanders and Hainaut	16 Alsace	27 Languedoc
6 Champagne	17 Franche-Comté	28 Dauphiné
7 Ile-de-France	18 Lyonnais	29 Comtat Venaissin
8 Maine	19 Bourbonnais	30 Provence
9 Anjou	20 Berry	31 Roussillon
10 Poitou	21 Marche	32 Foix
11 Touraine	22 Saintonge and Angoumois	33 Béarn

Map 1.1. French provinces in 1789. *Source:* P. M. Jones, *The Peasantry in the French Revolution,* Cambridge Univ. Press, 1988, 22.

The study as a whole resides uncomfortably within the years 1700 to 1860. Thus, one important period in the history of Old Regime water control is left out: the late sixteenth and early seventeenth centuries, when royal initiatives attempted to import drainage techniques from the Neth-

3

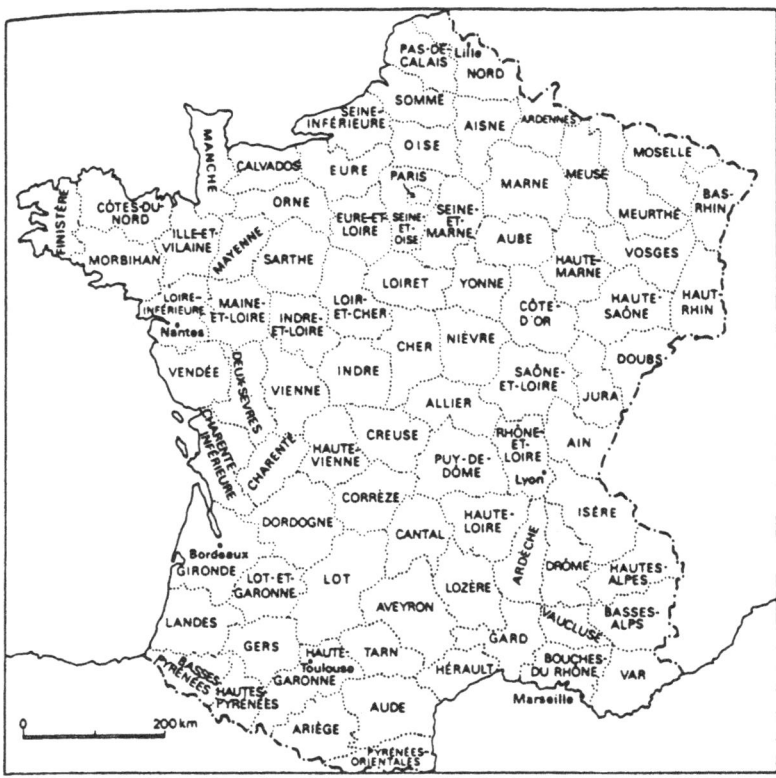

Map 1.2. The *départements* of France in 1790. *Source:* Donald Sutherland, *France, 1789–1815: Revolution and Counterrevolution,* New York: Oxford Univ. Press, 1986, 15.

erlands. This period was omitted primarily because of the scarcity of archival material on water control projects before 1700. I will argue in Chapter 4, however, that, to the extent that the evidence exists, it suggests that there were many similarities between seventeenth- and eighteenth-century water control.

If starting in 1700 was prompted by data considerations, stopping in 1860 stemmed from concerns with technological change. The pattern of investment in water control between 1700 and 1860 is simple: little or no investment before 1815 and much of it afterward. To explain this pattern, there are at least two alternatives to institutional change: relative price shifts and technical advances that could have increased the returns or decreased the costs to investment in water control. Relative price shifts between 1700 and 1860 will be investigated in Chapters 5, 6, and 7. While early nineteenth-century technology closely resembled that used before the Revolution, the pace of technical change quickened after 1860. For example, clay pipes came into regular use for drainage, and a number of other improvements further reduced the cost of water control. Since the techniques used after 1860 were not available in the eighteenth century, there can be no argument that those investments could have been realized in a similar fashion before 1789.

Although this book aims to analyze the impact of institutional change on investment in water control in France, the heart of the book is composed of two local studies: one of drainage in Normandy and one of irrigation in Provence. The number of local studies had to be limited because of the vast amount of archival material bearing on the impact of legal reforms. Indeed, as students of French history well know, legal institutions in Old Regime France were, despite the centralizing aims of the state, fundamentally regional. This problem was, if anything, exacerbated in the case of water control, where property rights often rested on medieval contracts or on custom. It was thus necessary to sift through the archival remains of a large number of water control projects to discover which institutions were most important. Another goal, no less intensive in archival requirements, was to examine quantitatively the impact of relative price changes. This further limited the number of regions examined. Studying the economic impact of the Revolution is difficult because of the absence of reliable statistical data before 1800 and because most of the qualitative evidence was created for polemical reasons. In an attempt to overcome these problems, I turn to a more microlevel source of information, which unfortunately yields regional rather than national conclusions. Although the regional conclusions may be quite strong, there is no obvious way to translate them into national aggregates. Thus, the national aspects of this study must be carried out with more qualitative data.

In order to best use the available evidence, the study includes explorations of explicit counterfactuals and analyses of detailed economic models, as well as an exposition of the historical particulars of each water control case. Historians may favor intuitive arguments over theoretical analyses, while economists may prefer a less microscopic level of inquiry. In this presentation I hope to satisfy the standards of both professions. Moreover, this book seeks both to persuade readers of the validity of its conclusions and to enable them to understand how they were reached. For all these reasons, many historical and technical details remain in the body of the text.

PART I

History and economics

2

The French Revolution and French economic history

The French Revolution was long thought to have been caused by social decay compounded by two acute economic crises.[1] The Revolution was seen as a convulsive transition from an outdated, feudal order to a new, capitalist order. The transition was precipitated by a half-century of increasing land scarcity and deteriorating standards of living topped by two consecutive harvest failures in 1788 and 1789. Adding to the trouble, the precapitalist system of taxation and borrowing no longer enabled the state to raise enough money to sustain itself.[2] These crises were thought to have heightened class antagonism since neither clerics nor nobles paid much in the way of taxes, and being wealthy they did not experience the pangs of hunger. The third estate alone bore the full brunt of high taxes and food shortages.

The interpretation of the Revolution as a social metamorphosis has been attacked on all sides, and even the crises that were reputed to have precipitated it have come under scrutiny. Starting with Alfred Cobban, historians have emphasized that the Revolution was less the replacement of a feudal elite and order by a bourgeois elite and order than a political realignment that divided all social classes between conservatives and radicals. The Revolution was carried out primarily by lawyers and administrators rather than by a capitalist bourgeoisie.[3] Furthermore, rural components of the Revolution – whether capitalist or conservative in their thrust – were viewed with deep suspicion by city dwellers.[4] As a result of this scholarship, distinctions between the old and the new order have become blurred, making the Old Regime's decay less tangible. Yet a more

---

[1] The bicentennial celebration of 1789 has focused the attention of historians on the French Revolution, and they have produced a number of excellent historiographic essays; see, e.g., Sutherland (1986, 15–48, 443–65), Doyle (1980), and Furet and Richet (1970).

[2] This interpretation is reviewed in Sutherland (1986, chap. 1).

[3] Cobban (1968, chap. 6); Schama (1989, pt. 1). For a more general presentation of eighteenth-century elites, see Chaussinand-Nogaret (1985).

[4] Sutherland (1986, 68–76, 93–114); Jones (1988).

general argument that institutions were out of step with the economy before 1789 has been largely uninvestigated.

The economic crisis that was thought to have broken the back of the Old Regime has also been elusive. The short-run evolution of living standards around 1789 has yet to be satisfactorily examined.[5] Nonetheless, it seems clear that food prices rose faster than daily wages after 1750, while prices of manufactured goods appear to have matched, or lagged behind, daily wages. Thus, the evolution of living standards depended both on changes in the composition of family budgets and on changes in employment. The quickening pace of economic activity may well have increased the regularity of employment. Indeed, local studies are full of evidence of growth in at-home textile production, increases in construction, investment in roads, and, above all, the spread of commerce.[6] Thus, employment may have risen sufficiently to maintain or increase living standards even if real daily wages fell between 1750 and 1789. All of this suggests that the existence of a late-eighteenth-century crisis in living standards is debatable at best.

In the extreme short run, there was indubitably a subsistence crisis in 1788–9. Yet short-term food shortages far worse than those of 1788–9 had been weathered before without precipitating a Revolution. The subsistence crisis may indeed have fueled the Revolution, because over the previous half-century expectations had changed and individuals had demanded state intervention to reduce the impact of food shortages. In a similar fashion the half-century of rising commodity prices should have increased the returns to agricultural investments like water control. Yet landowners were typically prevented from making such investments. During the eighteenth century pressure for the state to intervene in these matters mounted, yet no effective reform occurred. The Revolution occurred at a time when the French, poor and rich, were particularly disappointed by the lack of positive government intervention in the economy.

The budgetary crisis, unlike the social crisis and the harvest disaster, seemed incontrovertible; after all, the near bankruptcy of the state had played a key role in precipitating the Revolution.[7] However, recent work suggests that the state faced a short-term rather than a long-term crisis in 1787–9. Indeed, over the previous twenty years its deficits had not been unsustainable.[8] Yet a real problem existed. The French govern-

---

[5] David Weir (forthcoming) has begun to reexamine these issues.
[6] See, e.g., Perrot (1975), Lefebvre (1959), and for the South, Billioud (1956).
[7] Weir (1989); White (1989); David Weir and François Velde, "The Financial Market and Government Debt Policy in France, 1750–93" (Yale Univ., mimeo, 1990). See also Bien (1987) and David Bien, "Venality of Office on the Eve of the Revolution: The Stockbrokers" (Univ. of Michigan, mimeo, 1990).
[8] White (1989, 550–1).

ment's ability to raise taxes was limited; thus, its ability to repay loans was always in doubt. Future wars, administrators knew, would require ever greater resources; if France was to maintain its position in Europe, tax revenues would have to rise. Fiscal reform as much as fiscal crisis may have motivated the call of the assembly of the *notables* in 1787. What Louis XVI, or his ministers, had not anticipated is that attitudes had changed and the traditional solution, raising taxes without a change in the distribution of power, was no longer acceptable even to the aristocracy.

The Revolution's causes thus appear to lie as much with attitudes and expectations as with economic and social crises. There was a growing demand for reform between 1780 and 1789 over a host of different economic and social problems. The Revolution presented an opportunity to improve institutions significantly; thus, it is important to ascertain whether the French seized that opportunity.[9]

In the short run it appears that the Revolution did little to benefit the economy. In a recent article, Gilles Postel-Vinay focuses on the proximate consequences of the Revolution. He concludes that it was at best a subsidiary event in France's economic history with three consequences worthy of note – two positive and one negative.[10] The two processes that favored economic growth were land redistribution and a dramatic reduction in household debt due to the inflation of the *assignat* period (1793–5). Against these improvements Postel-Vinay tallies the enormous disruptions in trading and credit networks caused by Revolutionary upheaval. These networks had grown significantly in sophistication and size in the fifty years before 1789 and had to be reconstructed after 1815.

The Revolution often appears to have had even more disappointing economic consequences in the long run. No industrial revolution on the scale of that which occurred in Britain followed 1815; no dramatic improvements in agricultural productivity were witnessed either. In fact, the French economy failed to grow as rapidly as Britain's until at least as late as 1860.[11] The absence of rapid development in France may have been due to factors other than institutions. Many have argued that France's comparative advantage probably did not lie in iron and steel or in coal-based technologies. Moreover, as Robert Allen has found, most of the agricultural revolution in Britain was due to a rise in biological grain

---

[9] For the public finance effects, one can turn to the low interest rates paid by French governments after 1815; see Homer (1977). There were also no government defaults in the nineteenth century. For the effects of the Revolution on income and wealth, see Postel-Vinay (1989).

[10] Postel-Vinay (1989, 1045).

[11] The argument of French backwardness has been severely attacked in O'Brien and Keydor (1978). But the fact remains that per capita industrial output was lower in France than in Britain throughout the eighteenth and nineteenth centuries. See also Roehl (1976).

yields rather than institutions.[12] Much the same could be said about French agriculture in the nineteenth century. Hence, it is often argued that institutional change was not important to the economy. The only legacy of the Old Regime, and a persistent one at that, was a lack of urbanization, which George Grantham argues had much to do with the absence of technological change in agriculture.[13]

Yet measuring French economic performance by the standard of Britain's industrial revolution may be misleading. Given its endowment, France could not have followed Britain's industrial model; hence, it was likely to remain a highly agricultural country well into the nineteenth century. Agricultural performance would thus play a more important role in France than in England. Moreover, French geography made it unlikely that the country would benefit rapidly and uniformly from changes in biological grain yields. Thus, institutional change in agriculture may have been the most important agent of economic development for France. Thanks to the Revolution, the nineteenth century began under a rather different set of institutions than those that prevailed under the Old Regime. Leaving aside the well-known massive redistribution of land from the church to private individuals, many other property rights were dramatically rewritten.

The impact of the institutional reforms of the Revolution on agriculture has not received much detailed study. Indeed, most studies of the early-nineteenth-century economy have been conducted at the national level, often with census data from the 1840s without a comparable source from the 1780s.[14] Meanwhile, most Old Regime studies stopped their investigations in 1789. Nonetheless, the Revolution's consequences have been analyzed as part of a larger problem that involves the state of French agriculture in the eighteenth and nineteenth centuries.

The state of French agriculture has been a matter of debate for as long as French–British comparisons have been made. Most examinations of agriculture on both sides of the English Channel have resulted in an indictment of French farmers for the whole of the eighteenth and nineteenth centuries. This indictment antedates the analyses of modern historians, since French government officials in the 1770s already felt that rural Britain was both more productive and more prosperous than France. In contrast, they viewed French rural areas as mired in poverty and underemployment and fettered by the shackles of traditional production

[12] Allen and O'Grada (1988). For more detail, see Allen (forthcoming).
[13] Grantham (1978, 333).
[14] George Grantham's work (1978, 1989), for example, analyzes the performance of nineteenth-century French agriculture in the light of the first censuses of agriculture. But he can make little direct inference about the role of the Revolution.

techniques and property rights.[15] This view was echoed by the most famous of all eighteenth-century agricultural experts, Arthur Young.[16] Curiously, both Young and French royal administrators pinpointed superior institutional arrangements – be they enclosures, long leases, or large farms – as the source of Britain's advantage.[17]

Early commentators on the Revolution shared Young's conviction that institutions were paramount in the performance of agriculture. They felt that the Old Regime had stymied the transformation of rural France into a capitalist economy. The Revolution, they argued, opened up the countryside to economic growth. They concluded that most nineteenth-century differences between England and France were due to the pre-1789 experience and its persistent aftereffects. More recent scholarship has thrown some doubt on this assertion. As George Grantham has shown, growth was limited in the nineteenth century by institutional constraints and a lack of demand.[18] If anything, there was less innovation during the nineteenth century than previously thought. Open fields persisted until well into the twentieth century, and scattered strips were common in many regions until World War II, because unanimous consent was required for consolidation. Thus, it is clear that the Revolution did not eliminate all institutional barriers to agricultural development.

Modern scholars not only have reduced the achievements of nineteenth-century agriculture, but have favored market forces rather than institutions in explaining differences in development.[19] The emphasis on market forces pervades the work of historians who focus on eighteenth-century farmers. Although there is virtual agreement that, unlike Britain's enclosures legislation, Old Regime institutional reforms had little effect on the economy, the pattern of land rents suggests that productivity was rising rapidly after 1750 in a number of French regions.[20] Philip Hoffman's analysis of productivity levels in the Paris basin offers compelling evidence of rapid improvements after 1775, and the rise may have started even earlier.[21] Thus, Old Regime agriculture seems to have been rather more flexible and innovative than once thought.

French farmers between 1700 and 1860 did have the opportunity to make a number of marginal improvements but seem to have been unconcerned with wholesale changes in production systems. New crops, like vines in the South between 1750 and 1789 or sugarbeets after 1820 in the North, could be easily inserted into the traditional planting scheme;

---

[15] AN H¹ 1489–92. For an early analysis of reform attempts, see Bloch (1930).
[16] Young (1929). Curiously, Young's data do fully bear out his negative appreciation of French agriculture; see Allen and O'Grada (1988).
[17] Bloch (1929).   [18] Grantham (1980, 1989).
[19] Robert Allen, "Enclosures, Capitalist Agriculture and the Growth in Corn Yields in Early Modern England" (Univ. of British Columbia, mimeo, 1986).
[20] See Chapter 5.   [21] Hoffman (forthcoming).

hence, they did not require institutional change.[22] Even more traditional improvements, such as substituting wheat for rye or using better seed, required almost no change in overall technology. Such improvements suggest that the Revolution may not have played an important role in agricultural development. The Revolution's impact, however, may have been outside the traditional system.

Indeed, eighteenth-century reformers focused on improving marginal land. Their goals were to increase output through land clearing, drainage, irrigation, and the reduction of common rights to private property. They felt such improvements would lead to the achievement of their primary objective – a rapid increase in total output.[23] The goals of these Old Regime reformers may well have shaped the reforms enacted by the Revolution. Hence, independent of the achievements of the traditional system, the question remains whether the Revolution achieved the Old Regime goal of bringing more land into production.

The movement to improve land in the late eighteenth century took place over much of Europe.[24] Its principal objective was to carry out property rights reform over land to reduce the cost of improvements. Although water control was part of this movement, the most visible attacks on traditional property rights involved enclosures in Britain and attempts to break up common land in France, both of which have been intensively studied. Economists have focused on the increase in efficiency that new institutions may have brought. Historians, however, have highlighted the redistributional consequences of the laws that promoted enclosures and the breakup of common land.[25]

Both historians and economic historians have long believed that, in terms of raw output, enclosures were superior to open (scattered) fields, but they have differed in their understanding of why open fields persisted until the nineteenth century.[26] Donald McCloskey offers an economist's perspective on the causes of the creation, persistence, and decline of open fields.[27] He argues that under some circumstances, such as uninsurable uncertainty about harvests, farmers preferred to scatter their land. Indeed, scattering reduced the likelihood that a farmer's whole crop would

[22] For an example, see Postel-Vinay (1985).   [23] AN H¹ 1512.
[24] Grantham (1980, 517).
[25] For a discussion of the British literature, see Allen (forthcoming, chap. 1). In France, the investigation was started by Lefebvre (1959).
[26] The costs of open fields involved overgrazing on common land, overinvestment in grain production, insufficient investment in capital and livestock, neighborhood costs, and time lost commuting from plot to plot.
[27] McCloskey (1975).

suffer the same fate.[28] If risk declined sufficiently, then the efficiency losses of scattering would outweigh the value of insurance and land would become enclosed provided that institutions were sufficiently flexible. Thus economists have to a large extent argued that the timing of enclosures had little to do with regulation – discounting the importance of the famed enclosure acts. Rather, changes in relative prices increased the returns to enclosures and thus provided the stimuli for institutional change.

While granting a role to the secular evolution of relative prices, historians have emphasized the redistributional impact of enclosures.[29] Although enclosures were necessary to raise farm output, they also reduced the welfare of the majority of the rural population. The specific institutions that governed change, namely parliamentary acts, allowed large landowners to disregard the effect of enclosures on smallholders and tenants. Hence, from a social point of view institutions remained paramount. In his painstaking study of the English Midlands, Robert Allen concludes that the productivity gains from enclosures were rather small but that redistributional consequences were very significant and may have been the prime motivation for enclosures.[30]

The analysis of enclosures highlights two problems that also plagued water control throughout the eighteenth century. First, change is costly, and second, it involves redistribution. The property rights regime of the English open-field system was the result of centuries of interaction between lord and village, landowners and tenants. Thus, many individuals (landowners, tenants, and laborers) had potential claims on the benefits of enclosures. For large landowners, who sponsored the enclosure acts, the returns of enclosures depended on the mechanism whereby each individual's rights were taken into account. Had more attention been paid to smallholders, for example, the cost of enclosing land would undoubtedly have been higher. Smallholders, who had only a limited voice in the political arena, were dispossessed with little, if any, compensation. Enclosures were successful because Parliament was willing and able to impose a system of allocating property rights that kept the cost of enclosures low, with the result that many suffered from change.

The second movement toward property rights reform involved common land in France. In Old Regime France, even though the issues and ideologies were quite similar to those that prevailed in Britain, the politics of institutional change in agriculture were quite different. As in the case of British enclosures, property rights reformers aimed to increase agricultural output, while opponents of change focused on redistribu-

---

[28] Other authors have found economic incentives for scattering other than harvest risk. See Fenoaltea (1976).
[29] Hammond and Hammond (1948).   [30] See Allen (1982, forthcoming).

15

tional consequences. In France, however, the organizations that controlled reform had little interest in promoting change. There was no forum like the English Parliament for landowners to enact reforms that could disarm attempts to resist change. On the contrary, the judicial system in France offered numerous avenues for resistance to reform.

The reform movement was guided by the commitment of key French administrators to physiocracy, an ideology also espoused by advocates of enclosures in Britain. Physiocrats promoted the abolition of internal trade barriers and a strengthening of private property in agriculture.[31] Perhaps the most visible part of their reform program was an attack on common land. In almost every French village, some land was held in common. Such land was frequently in a state of well-defined use and poorly defined ownership. Both use and property rights were the result of centuries of interaction between a village and its seignior so that frequently ownership of the resource was unclear.[32] The reforms of common land promoted by the Physiocrats sought to divide the land among its various owners without necessarily offering compensation for lost use rights. As a result, reforms might have had a negative impact on village welfare. Historians have interpreted this attack on common land as a vast redistribution of wealth from villages into the greedy hands of urban dwellers and nobles.[33] Moreover, they have argued that reforms were successful only where they coincided with the aims of noble landowners and where villages were politically weak.

Even though the analyses of common rights in the eighteenth century have focused on redistribution, historians have also noted the prevalence of poorly used common land and marshes. Their work suggests that increasing land scarcity after 1700 actually exacerbated the inefficient use of common land. For example, Lefebvre in his studies of Orléans noted increased encroachment on forests, most of which were common land, in that region after 1740.[34] Because the ownership of forest land was unclear, it was improperly policed and the forest suffered encroachment. Because they were likely to be evicted, encroachers did little to preserve the quality of the land they cleared. Thus, as a result of uncertain property rights, valuable forest was transformed into mediocre pasture.

In his study of the North, *Les paysans du Nord*, Lefebvre also noted the disastrous effects of common access to peat bogs.[35] Peat bogs were an important resource in the North not only because they provided peat for heating, but also because they provided pasture. Again, because of uncertain ownership, no one had a long-term interest in the bogs. During

[31] Allen (forthcoming, chap. 1); Saint-Jacob (1960).   [32] Jones (1988, chap. 1).
[33] Saint-Jacob (1960, pt. 3). More recently, historians who have carefully combed the archives find evidence that a much larger proportion of the population took advantage of the reforms. See Clère (1988, chap. 2).
[34] Lefebvre (1962, Vol. 1, 36).   [35] Lefebvre (1930, 79–81, 224–230).

the late eighteenth century, peat was overmined, the bogs sank below the water level, and all pasture was lost.[36] Royal reforms that would have privatized peat bogs would no doubt have limited the damage.[37] Yet Lefebvre ignored these efficiency issues to focus on the other impact of royal reform: the dispossession of village use rights.

The vast literature that has followed Lefebvre's *Paysans du Nord* has similarly noted an array of problems associated with ill-defined property rights.[38] However, because scholars were concerned primarily with uncovering sources of tension that presaged the Revolution, that literature has focused on the redistributional aspects of common land reform. Reform, however, could not be equitable without facing extremely high costs because of the muddled nature of property rights over common land. As in the case of enclosures in Britain, administrators searched for a method whereby they could cut through the property rights morass that controlled common land. In France, as in Britain, the reforms favored owners (seigniors or landowners) rather than users (villagers or tenants) because they had much greater political influence. Yet the sometimes dramatic redistributional impact of reform proposals meant they were fiercely combated.

The reforms proposed to divide common land between seigniors, landowners, and tenants. Unlike in England, where all large landowners benefited from enclosures, in France some large landowners often nearly monopolized use rights at the expense of other landowners. These large landowners opposed property rights reform because it would reduce their access to pasture.[39] They were able to promote an ideology that legally and politically attacked the position of the champions of private property. The ideology of opponents of reform sought to legitimize the claim that use rights gave property rights to land. Although some historians have taken this ideology as an indication that peasant communities were harmonious, egalitarian, or non-market-oriented, the evidence is overwhelming that a few rich, market-oriented farmers often benefited disproportionately from customary rights.[40] Hence, the ideology of the sanctity of communal rights was little more than a screen designed to shield some villagers from the redistributional consequences of change.

The political, legal, and ideological opposition to reform was sufficient to deter reform except in a small number of provinces. It was not until

---

[36] Ibid. (81).

[37] One should note that within the royal administration not all were won over to private property rights and that some of the schemes proposed for dividing these marshes would have increased rather than decreased the rate of peat extraction. AN H[1] 1488.

[38] See, e.g., Jones (1988, 17) and Clère (1982).   [39] AN H[1] 1486, fl. 131.

[40] See AN H[1] 1486. For a detailed analysis, see Kathryn Norberg, "The Amoral Economy of Echalon: An Eighteenth Century Community from the Perspective of Its Seigniorial Court" (UCLA, mimeo, 1988).

the Revolution that questions of common land were addressed. By that time seigniors – one of the many parties in the debate – had lost their political power, so it was relatively easy to give all common property rights to villages. One should note, however, that under the Restoration (1815–39) the extent of villages' revolutionary appropriation of common land was tested. Law handbooks of the early nineteenth century described in detail the circumstances under which villages could maintain their claims to common land. There was nothing natural in the property rights regime that emerged after the Revolution because seigniors, for example, had been expropriated without compensation. Yet it was a regime that facilitated the transfer of fertile land into private hands – precisely the goal of eighteenth-century administrators.

Common rights were but one part of the central problem of the Old Regime: *privileges*. This complex system of entitlements extended through most sectors of the Old Regime economy. *Privileges* were granted by the king to individuals, groups, or corporations most frequently in exchange for ready cash. These grants gave their holders specific rights such as tax exemptions, production and sale monopolies, or even judicial positions. By 1700, few individuals in France were not beneficiaries of some *privilege*.[41] After 1700, economic expansion and the increasing prevalence of an ideology that valued merit over birth or wealth increased the economic and social costs of *privileges*.[42] In the long run, maintaining *privileges* would have reduced national output by increasing distortions in the economy and thus would have led to a lower tax base. In the short run, however, maintaining or increasing *privileges* increased tax revenue and ensured the loyalty of the holders of *privileges*. *Privileges* attacked by the Crown were most often defended by a traditionalist ideology that was not unlike the arguments that villages used to defend their use rights. Holders of *privileges* and villagers both argued that reform was an act of despotism that threatened the very fabric of French society. Yet historians have viewed this same ideology with favor when it was used to defend common rights and with scorn when it was marshaled in defense of other *privileges*. In neither case has a real effort been made to ascertain the extent of the gains from proposed reforms or even the extent of the redistributional consequences.

The ultimate failure of property rights reform under the Old Regime affected promoters of drainage and irrigation for different reasons. Drainers of marshes were directly involved with the reforms outlined earlier because most marshes were common land. Projectors of irrigation canals faced an even more complex problem of property rights because eminent domain and water rights were local *privileges*. In both cases, the Revolution achieved a complete redistribution of property rights and author-

---

[41] See, e.g., Bossenga (1988). [42] Chaussinand-Nogaret (1985, chap. 1).

ity. What was left for Napoleon and subsequent leaders was to create organizations that would further reduce the institutional costs of providing drainage and irrigation. But that task required no further wholesale redistribution of property rights and thus proceeded with little debate and opposition.

Now entering its third century of study, the Revolution is suffering from a rather bad press.[43] This negative assessment may derive from the increasingly frustrating search for an extraordinary crisis that should have preceded 1789. In the absence of a momentous crisis, the Old Regime should have been able to weather the challenge of 1789. Obviously it did not, leaving scholars with the uncomfortable conclusion that it failed due to an accident of history: the indecisive and incompetent Louis XVI. The human toll of the Revolution seems to have been an unnecessary price to pay for whatever institutional reforms occurred between 1789 and 1815.[44] Yet even the most optimistic proponents of the Old Regime find it hard to accept a Revolution due simply to error.

The Revolution, I wish to argue, was not an accident caused by the failure of Louis XVI or his ministers to understand the need for change. On the contrary, the negotiations between 1787 and 1789 suggest that the administration was intent on reform. The Revolution was an accident to the extent that neither the royal administration nor its opponents foresaw the events of 1789–95 when negotiating over taxes in 1787–8. Indeed, the early fiscal negotiations took place between the royal administration and an assembly of great aristocrats. Both groups had a fundamental stake in preserving the Old Regime. Both groups, however, wanted substantial modification of the French economic and political structures. What the central administration and a number of Revolutionary governments appear to have failed to realize was that they could not control the process of change. Thus, the Revolution began because there was a need for change, yet the circumstances of its beginnings may tell us little about what reforms were actually implemented. While testing such an interpretation of the causes of the Revolution is beyond the scope of this book, it does offer a valuable metric to evaluate the success and importance of the Revolution.

If we accept that the cause of the Revolution was a demand for reform, then it is fair to ask both how acute the problems that brought about

---

[43] Simon Schama's very popular *Citizens* (1989) is the clearest articulation of the negative perspective on the Revolution.

[44] One should note that the human price seems most frequently to ignore those soldiers – French or not – that died in the international struggles between 1791 and 1815. But their numbers overshadow any count of those who died during the Terreur or other episodes of Revolutionary violence.

that demand may have been and how successful the Revolution was at solving these problems. To be sure, undrained marshes and dry land did not drive the third estate into revolution. Yet water control does represent just the kind of agricultural problem that the Old Regime sought to address; hence, there could be no better proving ground for the impact of the institutional changes that followed 1789.

3

Institutions and economic growth

In the early Middle Ages, improved land was scarce and unimproved land abundant. As a result, lords and villages chose different property rights to govern these two types of land. Property rights to improved land were precisely assigned, giving rise to what we call private property. In contrast, it was not worthwhile to define property rights to unimproved land clearly, for enforcing such rights would have required monitoring unwarranted by the low value of the land. Thus, unimproved land fell under a regime of common rights. This system was efficient as long as it was possible to improve land when relative prices shifted. Yet poorly defined property rights would come to stand in the way of improvements in general and water control in particular. In order to earn a return on their land improvement schemes, water control projectors had to divide costs and benefits between themselves and the owners of the land. Carrying out such a division required an agreement on existing property rights (establishing who owned the marsh) as well as agreement on a rule that translated the original distribution of property rights into a share of the profits (benefits minus costs). Presumably, marshes could have been improved and still remain common. However, in the eighteenth and nineteenth centuries it was generally thought that development was incompatible with common rights. To complicate matters, by the year 1700, altering property rights from common to private required state approval.[1] Hence, the regulation of common land that had begun largely as private contracts between medieval lords and their serfs had become a public institution.

Economists and economic historians have agreed that private arrangements like bilateral contracts adjust easily to changes in the environment. Private arrangements seem to evolve to increase productivity or to adapt

[1] See Chapter 4 for details.

relative price shifts. For example, in his long-running study of the evolution of U.S. firms, Alfred Chandler concludes that new forms of business organization represented efficiency advances.[2] More important, private arrangements adapt quickly to new environments because the owners of a firm have both the incentive and authority to modify contractual arrangements when necessary.[3] In other words, Ronald Coase's contention that individuals will choose the most efficient institutions available is likely to hold.[4] Although in the short run new institutional arrangements may have negative distributional consequences, in the long run institutional improvements should improve productivity. As a result, change will benefit everyone in the firm.[5]

Our case concerns the set of public institutions known as the law. Unlike private institutions, law is not modified as a result of private agreements; rather, it changes as a result of political processes. Thus, we cannot assume that law will adapt to new environments rapidly or efficiently.

In economic history, the study of public institutions has been pioneered and promoted by Douglass North. North suggests that public institutions were similar to private arrangements. Following Coase's seminal article on the nature of the firm, North argues that the positive function of institutions is to minimize transaction costs – that is, the costs of specifying, trading, and enforcing property rights.[6] Within this general principle, North has offered a variety of propositions concerning the extent to which institutions adapt to changes in the environment. In an early work, *Institutional Change and American Economic Growth*, North and Lance Davis suggest that institutions are continually adjusted to reduce transaction costs.[7] In this view, changes are stimulated by political entrepreneurs who earn a return from the increase in efficiency. Thus, public institutions respond to changes in the environment in a manner similar to the rules that control Chandler's firms, and economic growth follows.

Yet economists quickly realized that the alignment of control and benefits that one finds in private arrangements is not found in the case of the law. Indeed, those individuals or organizations that control the law will often have incentives to do something other than maximize national income. In *Structure and Change in Economic History*, North revises his

---

[2] This perspective is most clearly articulated in Chandler (1977).
[3] We should note, however, that the firm may be controlled by managers whose incentives are quite different from those of the owners. Hence, full efficiency is not guaranteed even under private arrangements.
[4] Coase (1937).
[5] For a counterexample, however, see Robert Allen's discussion of the effect of enclosures on employment in the English Midlands. Enclosures reduced the demand for labor in a major sector of the economy, agriculture. Allen (forthcoming, chap. 10) argues that they were sufficiently widespread to affect wages adversely.
[6] Coase (1937, 338); North (1981, 5).   [7] Davis and North (1971, pt. 1).

early views on institutional change. In this later work, he explicitly considers the political economy questions of institutional change. He argues that, while the state's ability to alter institutions can lead to increases in national output, it can also offer opportunities to reallocate wealth and income across groups.[8] According to North, since individuals care more about their income and wealth (redistribution) than about national output (efficiency), state reforms will be the outcome of political pressures, which will be concerned mainly with redistribution.[9] Political entrepreneurs and individuals might, as a result, devote more effort to reallocating existing wealth than to increasing efficiency – a process known as rent seeking.

The importance of redistribution in setting off opposition to reform under the Old Regime can be highlighted by an examination of the impact on a village of draining a marsh. Before drainage, villagers had free use rights to the marsh. Moreover, an acre of undrained marsh was worth nearly half that of an acre of drained marsh. In such circumstances, villagers would have had good reason to oppose drainage without compensation for their use rights. Indeed, drainage would deprive them of at least as many resources as it would create. Yet for some villagers, the consequences of such a proposal were even worse. Because access to marshes was frequently unequal before drainage, a few individuals were bound to lose a great deal of cheap pasture. Conversely, if the government adopted a scheme that compensated villages for lost use rights, it effectively reduced the profits of drainage projectors and seigniors. All parties, it seems, would be willing to expend resources in order to influence the content of the law that governed drainage. These resources were expended in the political sphere. Thus, it is possible that the reform proposals were intended to steer the gains of drainage toward politically important groups.[10]

More generally, given the inefficiencies of state reform, some economists have invoked a rule first proposed by Ronald Coase: If transaction costs are zero, then individuals will arrive at an efficient allocation of resources independent of the original assignment of property rights.[11] In this view, the state's most important role is to define and enforce property rights (minimize transaction costs), leaving private individuals to take care of the rest.[12] Yet the state can never define property rights perfectly, nor does it refrain from regulating economic activity, making transaction

---

[8] North (1981, chaps. 3 and 11).
[9] For an economist's perspective, see Tullock (1967).
[10] The role of the state in economic development is the subject of a vigorous debate in the field of economic development that has many parallels to the same debate in economic history. For a recent example, see the issue of the *Journal of Economic Perspectives* (Summer 1990) partially devoted to this debate.
[11] Coase (1960).   [12] Demsetz (1964, 1966).

costs inescapable. As a result, the economy's performance depends on both the magnitude of transaction costs and the original assignment of property rights. The impact of transaction costs will be mitigated by the efforts of individuals who will attempt to contract around inefficient laws (e.g., common rights). Yet if it is very difficult to circumvent or change a rule, then its inefficiency consequences can be very high. One must, therefore, ascertain the costs of changing rules – if they are very high, laws can have large negative effects on efficiency.

In the specific case of water control, two factors weighed most heavily against a low-cost private alternative to law. First, water control faced significant problems of coordination that were not easily solved privately. As noted in Chapter 1, coordination was important because the number of individuals affected by a water control project was often large. Moreover, since some water control projects were local public goods (as in the case of drainage), it was difficult to keep even those individuals who did not want to participate in the costs from reaping the benefits of a project.[13] As we shall see, these problems made private contracting quite difficult. Hence, the diffuse nature of property rights led to transaction costs that were high relative to what they would have been with more concentrated ownership.

Second, the very profitability of water control was a source of transaction costs. The principal problem was the division of surpluses created by water control projects – drained land in the North or irrigation water in the South. Many participants could halt water control projects by refusing to sign the contract dividing the surplus. As a result, projectors, villages, and seigniors faced a significant bargaining problem.[14] That bargaining problem had nothing to do with the social value of the project but everything to do with allocating its profits. Most economists assume optimistically that bargaining problems get solved somehow. Yet there is increasing evidence, even in the theoretical literature, that bargaining failures occur even when the surplus to be divided is large.[15]

In the case of water control the cause of bargaining failures was simple; individuals wanted the largest possible share of the surplus for themselves. In the absence of a rigid rule for surplus division, which would

[13] Similar problems were caused by property rights regimes that governed Texas oil fields. See Libecap and Wiggins (1985) and Wiggins and Libecap (1985).
[14] Private contracting is by definition voluntary, so it requires unanimity.
[15] George Mailath and Andrew Postlewaite, "Workers Versus Firms: Bargaining Over A Firm's Value" (Univ. of Pennsylvania, mimeo, 1989); David Porter and Jean-Laurent Rosenthal, "Bargaining Costs and Failures in the Sealed-Bid Double Auction" (Caltech, mimeo, 1990). These papers provide theoretical justification for bargaining failures that do not depend on the possibility that an individual will be made worse off. Another facet of the bargaining problem that is frequently assumed away is that courts can be used to enforce bargains; yet in the case of water control, that was far from true.

have amounted to a complete assignment of property rights, individuals had to find a division contract privately. Given the requirement of unanimous consent, individuals had an incentive to act strategically by demanding a large share of the surplus in the hope that others would reduce their demands. Thus, even in the case of certainty about the value of an institutional change to each individual, private solutions may not have been free of transaction costs. And in the case of marshes, common land, and irrigation water, property rights were most frequently uncertain, a fact that further complicated private contracting.[16]

The fact that development required the unanimous consent of all concerned had a dramatic impact that can be illustrated by a simple example. Suppose a wealthy projector understood the inability of a village as a whole to agree on drainage because of a unanimity rule. One might presume that the wealthy projector could have bought out the village, drained the marsh, and sold the land back to the villagers at a large profit. This scheme, however, was impracticable, for unanimity created a hold-out problem in which each villager had an incentive to be the last villager to sell out to the wealthy projector. The villager who sold out last would be, in effect, a monopolist and would be able to extract far more than the market value of his land from the projector. As a result, no one would want to sell and no drainage would occur.[17]

The state could have presented an easy alternative to this dismal state of affairs by changing property rights. Assume, for example, that owning two-thirds of the land in a village had been sufficient to decide on drainage of the marsh, or that the agreement of half of the heads of households in a village had been sufficient to decide drainage.[18] In these cases, the state could have played a background role in drainage, because individuals would have had an incentive to carry out the improvement by themselves. In both scenarios, all marshes with a positive social value would have been drained. In addition, a wealthy projector's scheme to buy up two-thirds of a village's land would have succeeded because there would have been little incentive for landowners to hold out. One should note, however, that these schemes would have been unlikely to increase every villager's wealth. Thus, the problem of redistribution remained.

Both the unanimity and two-thirds majority rules represent complete property rights assignments, yet one features smaller transaction costs

---

[16] See Chapter 6 for details.

[17] The same problem plagues contemporary urban redevelopment efforts. In order to rebuild certain areas, developers must assemble large tracts of land. Small landowners have an incentive to hold up the project for their personal gain. This problem is frequently addressed by giving a semipublic agency rights of eminent domain.

[18] The former scheme was in fact proposed in France in the eighteenth century, and it was by and large the scheme of British enclosures. The latter scheme was implemented after the Revolution.

25

than the other. This highlights the fact that the state has an important role in economic development because, like Chandler's entrepreneurs, it must find institutions that foster growth. If the state fails to promote property rights arrangements that reduce transaction costs, individuals will attempt to overcome the problem. Their success or failure will depend intimately on the structure of property rights over the resources that must be reallocated, on the strategic opportunities presented by the bargaining problem that arises, and finally on the extent to which the general structure of property rights – the law – enables individuals to circumvent specific arrangements.

The premise that the original assignment of property rights matters is not novel, but it does raise important methodological issues for the study of institutional change. In particular, it means that a great deal of historical detail is required. One might attempt to comprehend the general impact of law on water control in eighteenth-century France through a national study. Yet only by studying specific regions can one grasp the precise arrangements that governed water control projects. Furthermore, only by investigating specific projects can one estimate the social returns to drainage and the redistributional consequences of improvements.

Since public institutions are chosen in the political arena, the problem of institutional change is a problem of political economy.[19] Confronting institutional change as a political economy problem allows us to uncover some of the forces that shaped property rights reform. Let us first turn to the question of institutional choice before 1789 and after 1795. Outside of this Revolutionary period, legal changes were only marginal, and it is possible to use the political economy framework pioneered by Stigler, Pelzman, and Olson to weigh the importance of interest groups, class concerns, political power, and organizations in shaping new institutions.[20]

Because most of the evidence relating to the process of legal reform is qualitative, it seems unnecessary to build an analytical model. Rather, let us focus on the important assumptions behind models of political economy to extract a few crucial conclusions. While the political economy framework was built to investigate democratic systems, water control problems in France were primarily a bureaucratic affair. Hence, some adjustment to the models is necessary. In most political economy models, groups influence policy makers' decisions because they can threaten them with the loss of their position. For example, individuals tend to vote for

[19] Bates (1981) uses this perspective to illuminate how property rights were organized in colonial Africa.
[20] Stigler (1971); Pelzman (1976); Olson (1971). For an application, see Gilligan, Marshall, and Weingast (1989).

representatives who enact legislation that they favor and against those who promote policies they dislike. Interest groups who can influence a large number of individuals by publicizing an official's decision are therefore powerful.[21] Of course, a small interest group may offset the limitation of its size by providing resources other than votes (such as money). A similar logic prevailed in the case of water control, both before and after the Revolution. The basic argument is presented first; some examples to buttress the assumptions follow.

Under the Old Regime, the political economy of water control featured three sets of actors: the state, local administrators, and villagers. I will discuss these in turn. Let us begin by assuming that eighteenth-century French people were motivated primarily by material considerations. To be sure, some individuals championed water control for ideological reasons, yet it seems clear that most of the important actors who favored reform stood to gain from change. In other words, when it came to water control reform, individuals appear to have taken ideological positions that were congruent with the personal loss or gain that a change in property rights might entail.

The most important actor in the model was the royal government, which, for convenience, I will call the state. We can assume that the Old Regime had three different concerns. First and foremost were monetary concerns; that is, the state needed funds for military purposes. Second, the state cared about economic growth. Third, the state preferred to avoid internal dissension. Because the state had multiple objectives, it had to make trade-offs between reform – economic growth – and revenue or protest. In this scheme, if monetary concerns were very pressing, the state would have been willing to renegotiate, or put off, a reform in order to raise revenue or reduce protest. Thus, the state may have been willing to sacrifice long-term growth in order to avoid protest or to raise funds.

Consider now the second set of actors: local administrators. Concerned with their personal gains and losses from reforms, they weighed the administrative reward they might receive from successfully implementing reform against the effort that change required. Indeed, local administrators were, to a large extent, judged on their ability to carry out a host of tasks other than reform. If promoting reform made it difficult to carry out routine tasks because the local elites were opposed to change, then the administrators would probably have preferred to avoid change. Thus, they would have promoted only reforms with little redistributive impact.

At the bottom of the political pyramid stood the individuals who controlled water and land – the villagers. This group can be subdivided into four smaller groups: lords, large landowners, small landowners, and

---

[21] Olson (1971, 53–6).

farmers, with each playing a role in deciding the fate of reform.[22] Each group's claims on common land or on water was based on different interpretations of customary rights and local *privileges*. So reforms that altered customary rights and *privileges* would be assessed differently by each group.

While the individuals within village groups tended to have the same attitude toward reform, political change was influenced by the strength of coalitions. We can define a coalition as a collection of individuals who try to alter the content of a reform proposal. Because of the economic nature of institutional change, coalitions expended resources to affect the content of reform. Given a reform proposal, we expect that two coalitions formed, one in favor of change and one against.[23] Coalition formation depended on the existence of organizations that structured political activity. Where national organizations could direct political behavior, coalitions were likely to have been broad. In this case, individuals may not actually have acted in a fashion consistent with their short-term self-interest because the coalition could promise long-term benefits. If there were no such organizations, however, we would expect coalitions to have been local and issue specific.[24]

Now let us consider what might have occurred when a reform proposal to improve the structure of property rights was put forward. Given the existence of coalitions that lobbied to affect the content of legislation, the state had some incentive to alter the proposed reform in return for decreased protest, increased political support, or a direct monetary reward. In short, the state was willing to renegotiate if the monetary return was sufficiently high relative to the sacrifice in efficiency. More important, the state would have required a much larger compensation in terms of revenue to change a national law than a regional one because the total sacrifice in efficiency was much larger. Since groups had limited resources, it was easier for local coalitions to affect local or regional reforms than national ones because the state would require a smaller payment. Thus, the more local the change in property rights, the more likely that the state would renegotiate with small coalitions.[25]

Abstracting from the Revolutionary period, the political economy principles already outlined can be used to examine both the Old Regime's failure to enact meaningful reform in water control and the successes of

---

[22] Of course, it is possible that one individual was all of these at once. The categories are designed to enhance our understanding of each individual's motivation rather than to mimic reality.

[23] To be sure, the success of these coalitions depended on organizations that would force members to participate in the expenditures necessary to effect reform.

[24] Olson (1971, 43–52).

[25] One should note that, if individuals knew that the state was very willing to renegotiate, they would have opposed reform more strongly than if they thought the state was bent on reform.

post-Revolutionary governments. These principles cannot be tested in the rigorous fashion that cliometric economic history would require because of the scarcity of quantitative evidence. Yet they provide a method for organizing the evidence about what forces shaped reform before and after the Revolution.

There is considerable evidence that the Old Regime's objectives were consistent with our model. In eighteenth-century France, government officials responded to interest-group pressure.[26] Officials appear to have taken most seriously two kinds of threat: (1) revolts or disturbances and (2) the withholding of services or funds. Large groups, like poor villagers, resorted to protest to alter royal decisions. Because the army was a blunt instrument, restoring order was very costly to the state. Hence, the government preferred policies that avoided protest.[27] Meanwhile, smaller groups like wealthy *privilege* holders had positive as well as negative means of influencing the state. Wealthy *privilege* holders were frequently government agents who could effectively threaten to shut down the bureaucracy.[28] Perhaps more important, they were rich and could influence policy by refusing to provide loans to the government. While the state could force the wealthy to furnish these loans, this again was a costly way to raise money. Clearly, the state preferred policies that did not threaten existing *privileges*. Thus, not unlike a democracy, the Old Regime faced costs when promoting institutional change. Both massive protest and reluctant financial participation had severe effects on government administration – effects the Crown preferred to avoid.

The *intendants'* reaction to proposals to reform enclosure legislation in the 1770s illustrates the role of local administrators. The *intendants* were in charge of local police, poor relief, extraordinary justice, and tax collection. The example of enclosure legislation of the 1770s, a major episode in French agricultural history, highlights important Old Regime political constraints.[29] Parisian bureaucrats who supported enclosure legislation relied almost exclusively on reports from *intendants* for their information. Most of these reports suggested that reform would produce few benefits in their provinces. One theme recurs in almost all the corre-

---

[26] This is consistent with the findings of historians on the importance of local elites in shaping policy. If anything, historians overemphasized the power of one group, the aristocracy, over others. See Saint-Jacob (1960, 145).

[27] The best example of the bluntness of army intervention comes from the late seventeenth century. The policies of Louis XIV against Protestants led to revolt in the Cevennes (a region in south central France). The army's subsequent intervention led to the complete destruction of the local economy.

[28] That occurred to a large extent when the state attempted to reform the judicial system in the 1770s. See Mousnier (1979, Vol. 2, 645–53).

[29] See Bloch (1929) and Hoffman (1988, 241–64) for the two chronological extremes of this literature. The bulk of the correspondence between Paris and the provinces is in AN H^1 1486.

spondence from those *intendants* who opposed reform. They argued that enclosure bills would upset the balance of village communities and, more likely than not, increase poverty and unrest.[30] But while the poor might lose from enclosures, *intendants* saw the large landowners who effectively controlled villages as the major source of opposition to enclosures. Thus, the *intendants* were loath to support enclosure edicts, which they believed would lead to serious conflicts with local elites and might increase poverty. In other words, they opposed legislation that made their administrative tasks more difficult.[31] In the few areas where reform bills were enacted, local elites were able to modify the content of the legislation significantly.[32] Presumably, these changes allowed the *intendants* to maintain relative peace in their administrative regions.

Now let us turn to the third set of actors: villagers. Resistance or support for specific royal reforms cut across class and social lines, creating groups that under different circumstances might have been in opposition. Agricultural reforms divided villages in a number of ways. For example, resistance to enclosure reform sometimes grouped seigniors with the poorest villagers because together they stood to lose the most from change.[33] While seigniors were threatened with the loss of fallow pasture, their poorer neighbors faced the loss of gleaning rights. The same two groups were allied in support of common land reform because they stood to receive more land than they ordinarily controlled. They were opposed by *laboureurs* – large farmers who monopolized common pasture.[34] In yet other instances, like drainage legislation, seigniors fought alone against most of the villagers.[35] Finally, we know of at least one instance in which a seignior allied himself with the largest landowners of a village in an unsuccessful attempt to drain a marsh in Normandy.[36]

Water control coalitions were issue specific and locally based because the state was forced to enact reform at the regional, or even village, level. Indeed, the state was forced to take into account regional organizations like *parlements* or institutions like *privileges*. Since most reforms affected only one *privilege* (access to a marsh or water from one irrigation canal), coalitions formed primarily in reaction to very specific laws, and in the case of water control, individuals maximized their short-run gains. Similarly, there were no national coalitions because there were no national political organizations. To be sure, the *parlements* could claim to be both judicial bodies and political organizations, but they were regionally based and represented only certain interests within a region. In short, the Old

---

[30] In the case of drainage, protest occurred frequently and was more than once sufficient to discourage a project. AN H[1] 1486–9 contains many examples of projects stopped by protest.

[31] AN H[1] 1486.   [32] Hoffman (1988, 254).

[33] AN H[1] 1486.   [34] Apparently this was the case in Alsace. AN H[1] 1486, fl. 131.

[35] See Chapter 6.   [36] AN H[1] 1489, marsh of Ecramville, 1784–9.

Regime state faced small coalitions whose members felt the gains or losses of reform directly. Moreover, because the royal government was required to enact local legislation, it had strong incentives to renegotiate.

Let us examine in greater detail the evidence of the state's willingness to renegotiate.[37] Reforms in favor of increased drainage, irrigation, or, for that matter, agricultural investment were never widely applied.[38] In fact, most reforms were either abandoned or modified time and time again in an effort by the government to reduce protest or accommodate politically powerful individuals. Thus, though many drainage projectors were able to secure special legislation to improve land, village groups in turn were able to get the same special legislation modified to increase the benefits they received from improvement. Coalitions in favor of development were simply unable to satisfy the demands of those opposed to change. Yet despite these problems, the Crown was unwilling to put a stop to the renegotiation process because water control presented a thorny fiscal dilemma in the eighteenth century. Specifically, in the short run, improvements had the undesirable effects (for the Crown at least) of either increasing state outlays or lowering tax revenues. State outlays would increase because it was clear that the administration would have to devote considerable effort to impose these changes on the countryside, given their inevitable redistributional consequences. More problematic for the French Treasury were the effects of reform on taxes. In the short run, at least, most proposals would adversely affect the tax base. Thus, the government promoted reform only when state finances were relatively healthy. At other times, it preferred to reallocate property rights simply to raise revenue.[39]

The ambivalent attitude of Old Regime administrators toward reform further raised the costs of institutional change in agriculture.[40] To be

---

[37] A perverse form of renegotiating prevailed under the Old Regime. When the Crown was strapped for cash, it threatened to reform some *privileges* in order to ransom their holders. Nowhere was this more in evidence than with venal officeholders. At times, the Crown threatened to increase the number of holders of a certain office to increase administrative efficiency. The proposed rise in the number of holders would not be compensated for with an increase in the revenues of that office, thereby reducing the income of incumbents. Nonetheless, someone was most often willing to buy the new position from the Crown, because it provided a share of the revenues to that office. Yet this reform was renegotiable, and incumbent officeholders were frequently able to pay off the Crown and avoid diluting the revenues of their functions. If holding up officeholders for cash is not the hallmark of an efficient administrative process, it highlights the fact that under the Old Regime everything was negotiable.

[38] See, e.g., Hoffman (1988); for further discussion, see Chapter 4.

[39] See Chapter 6.

[40] One should note that Bertin and most other high-level administrators were large landowners and seigniors who stood to gain directly from these reforms. See Forster (1980, 77). Other administrators, however, surely stood to lose from these reforms.

31

sure, some villagers would have protested against water control reform simply because it threatened to reallocate valuable property rights, yet the state's willingness to renegotiate heightened opposition. Whenever the state attempted to enact agricultural reforms, it was clear, even to villagers, that these reforms would be withdrawn if it was demonstrated that the cost of imposing them was very high. Thus, opposition was a fruitful strategy, a fact that in turn raised individuals' willingness to resist. As a result, the cost of reform was much higher than it would have been in the absence of renegotiation.[41]

This model also highlights which constraints were eased by the Revolution's reforms and which factors still limited the process of change. In many ways it was easier to reform property rights after the Revolution. First, the state had the ability to raise taxes without tampering with property rights, so that the only constraint on reform was political support. Second, the state passed national laws, and there simply existed no local organization that could exert significant pressure on the central government.

Nonetheless, the concern for political support hindered change in agriculture during the nineteenth century, and as a result many property rights that stood in the way of development remained unchanged. For example, consolidation of holdings would have necessitated the reapportionment of privately owned land. Such widespread change was very unpopular, so the system of requiring the unanimous agreement of all landowners remained in force until the twentieth century.[42] Enclosures, however, were less fractious, and individual landowners were free to chose whether to fence their land or leave it open to fallow pasture.

The rule that was adopted after 1789 for dividing drained common land also suggests that political support was an important consideration in framing the law. The law required that a proposal for draining common land receive the majority of votes in a village council, thereby allowing large landowners who controlled the councils to decide on improvements.[43] This was important to the government because large landowners were a crucial political force. The law on drainage also allowed each head of household to buy an equal share of the drainage project, guaranteeing a widespread diffusion of early benefits. In addition, because the land was returned to private property, large landowners, who had been the biggest users of the commons in the eighteenth century, could then reconstitute large pastures if these were profitable.[44] In short, draining

---

[41] See Chapters 6 and 7.   [42] See Grantham (1980, 517).

[43] Thus, not all common land was divided and not all marshes were drained, especially in areas where large landowners continued to monopolize access to common land. See Norberg (1988, 265–86).

[44] The market for drained marshland seems to have opened before the marshes were actually divided or drained. Individuals in Troarn bought and sold their shares of

common land under these rules would guarantee that it was supported by a very wide constituency. Whatever the advantage of these rules, they were used successfully, because villagers knew that alternatives were not available. The state was no longer willing to engage in different negotiations for each marsh to be drained.

While this political economy model sheds some light on institutional change before and after the Revolution, it leaves us without a guide to the period from 1789 to 1795. In May 1789, the relative power of individuals and groups in France changed forever. A whole new set of individuals, with new ideological commitments and power bases, gained control of the state. Between 1789 and 1795, concerns for the survival of the Revolution also altered the equations of the political economy calculus. Armed confrontation with both foreign enemies and domestic insurgents changed administrators' concerns, which in turn may have either lowered or raised the cost of reform. All these changes occurred as the Revolution brought on fundamental institutional reforms. Each of the dramatic changes in the political arena may explain the increased interest in property rights reform that followed 1789. As a result, we have many possible explanations for what occurred during the Revolution and not enough information to distinguish accurately which causes were in fact important to institutional change. Given these limitations, when studying specific regions and specific institutional changes, it seems appropriate to take the Revolution and its reforms as given.

———

A final task remains – finding a framework for understanding why institutional change occurred during the Revolution and not before. As noted earlier, revolutions are difficult to model in a political economy setting because they do not produce marginal changes. Yet an alternative set of economic models provides insights as to why the Revolution could accomplish things that proved too difficult for Old Regime and nineteenth-century governments alike. Indeed, following the lead of Brian Arthur and Paul David, one can describe changes in public institutions as a path-dependent process.[45] That is, fundamental institutional change is a process that has very high fixed costs. As a result, fundamental change will not occur incrementally to adjust to the changing environment. Rather, change will come only after a significant departure is made from the optimal institutional structure. Moreover, at most points in time, the property rights structure that prevails will be highly dependent on the previous property rights structure, even if such a structure was inefficient.

the communal marsh a full year before it was divided and drained; AD Calvados, 8 E 25280.
[45] Arthur (1989); David (1985). For a discussion of the relationship between path dependence and institutional change, see North (1990).

Let us consider legal change within this framework. Laws are in effect part of a larger system, which has both basic and marginal components. Basic components include political organizations and their institutions, or the legal system. Marginal components consist of economic organizations and institutions (laws, private contracts, and judicial interpretations). While it is possible to modify marginal components somewhat, such modification is constrained by the basic structures of the institutional system. Changing the basic structure is very costly because it leaves all property rights undefined.[46]

For example, it was possible to reform property rights to marshes before the Revolution, but the political process made it very difficult and thus change was limited. Similarly, the cost and returns to a specific enclosure proposal depended on the distribution of political power. In some political systems, enclosures are carried out easily, in others at great cost, and in others still, it is impossible. In a sense, modifying marginal components allows society to adapt only partially to changes in the environment. Thus, as long as a given set of basic components prevails, change will be limited.

Yet systemic institutional change will occur infrequently because it is very costly. Indeed, only a dramatic departure from the optimum should make individuals desire systemic change. Once systemic change occurs – through the creation of a national assembly, for example – then the cost of modifying marginal components can be dramatically reduced.[47] We see this principle operating in the fact that the Revolution's political reforms broke the deadlock that had stifled both drainage and irrigation under the Old Regime.

One of the strongest arguments for viewing institutions from this perspective comes from the vast literature stressing the cost of uncertain property rights.[48] This literature emphasizes the point that individuals will make productive investments only if they are guaranteed a sufficiently large share of the fruits of those investments. If property rights are constantly threatened, individuals will be unwilling to invest in pro-

---

[46] The classic case of path dependence as articulated by Brian and David rests on increasing returns to scale. Since the law is by nature monopolistic, it shares many of the characteristics of an increasing return to scale industry. An alternative cause of path dependence is articulated by North (1990, chap. 11). He focuses on network externalities. For the purpose of this chapter, it suffices to recognize that either source of path dependence produces the same result: Changing the basic institutions is extremely costly. Thus, basic change will occur only under extreme circumstances.

[47] It should be acknowledged that, if rent seeking leads to inefficient institutional change in the traditional political models, the same can occur in a path-dependent process. Rent-seeking individuals may seek a systemic change because of the likelihood that they will benefit from the redistribution that follows. Hence, revolutions may occur too soon.

[48] Coase (1960); Demsetz (1964).

ductive assets, and the economy will fare poorly as a result. Even if changes in property rights are both rapid and definitive, the transition period from one system of institutions to another may be costly. Individuals will devote considerable effort to avoiding the distributional consequences of change. Such effort is, from a social perspective, unproductive since it involves redistribution and not production. Moreover, faced with systemic change, groups and individuals will have extremely strong incentives to fight that change through revolt or protest. Resistance will force the political authorities to assert their power during the transition – a process that in the Revolution proved very costly.

The persistence into the eighteenth century of many medieval property rights, despite the cost they imposed on the economy, lends support to the argument that institutional change is path dependent.[49] During the medieval era, the arrangements that led to the creation of common land may have actually been relatively efficient. Over the next few centuries, nothing challenged the basic structure of these property rights even though a number of marginal changes were made. For example, some of the property rights to common land were ignored, while some were transferred from seigniors to villages or between villages and private individuals. By the eighteenth century, "feudal" property rights created a complex and uncertain set of claims to resources. Because of accompanying relative price shifts (most notably an increase in the scarcity of land), the value of the resources itself became sufficiently high to require a more precise assignment of rights.

Three avenues were available to rectify this situation: Two were possible under the Old Regime, and one required a revolution. Before 1789, the traditional method for defining property rights was a slow judicial determination of the validity of claims made by opponents to improvements – an extremely expensive procedure that frequently led to greater rigidity of property rights. This judicial approach was actually attempted, and because the initiators were most often nobles, it has often been termed the "seigniorial reaction." Although, as we shall see, some property rights were clarified by the judicial procedure, the costs of such a process discouraged most projectors. An alternative Old Regime solution was property rights reform, and that proved too costly as well. Unfortunately for Physiocrats and other champions of agricultural development, marshes and irrigation canals were not important enough to create the condition of systemic social change. Hence, reform would be possible only as part of a more general process of change whose causes lay outside of agriculture – a revolution. In this perspective, the Revolution represented an opportunity for institutional change in areas where

---

[49] Some of these costs are the enormous legal expenses borne by lords and villages during the seigniorial reaction when feudal claims were reasserted. See Chapter 8.

the usual political calculus favored the status quo. This is not to say that the Old Regime was beyond salvaging; indeed, it could have been reformed. But in the area of institutional change, the absolutist monarchy was at a disadvantage. The contractual nature of the Crown's relationships with groups and individuals made change difficult because the state lacked the authority to change arrangements unilaterally.[50] Every new law required a complex set of bargains with a large number of groups. The Revolution voided these requirements and ushered in both a new set of basic institutions and a new technology for marginal changes.

---

[50] The absolutist monarchy tried to assert its authority over contracts, yet it is not clear whether it succeeded. See Mousnier (1979, Vol. 1, 659).

PART II

Drainage and irrigation

4

A survey of water control projects

Attempts to drain marshes or build irrigation canals are recurring themes in the agricultural history of most French regions. These attempts were subject to the scrutiny of a number of organizations. Indeed, under the Old Regime, royal administrators had authority over changes in water use, while local authorities controlled changes in customary practices or eminent domain, and finally, judicial authorities resolved conflicts between projectors and property rights owners.[1] In the nineteenth century, oversight was concentrated in the agents of the Ministry of the Interior, but these agents lavished at least as much attention on water control projects as had Old Regime administrators. The important role played by government officials in water control both before and after 1789 accounts for the abundant archival record covering this branch of agriculture. This record reveals the fate of many French drainage and irrigation projects between 1700 and 1860. In the case of drainage, the archival evidence is supplemented by a large number of local studies – many of which were written in the first half of the twentieth century. These studies were generally concerned with the institutional causes of the Revolution, and they investigated in great detail the legal aspects of property rights to common land and marshes. They thus contain a wealth of useful information about the scope and fate of water control projects under the Old Regime.

This material suggests that until the seventeenth century land reclamation through drainage and irrigation followed relative price movements. In other words, when land was relatively scarce, marshes were drained and irrigation canals were built. Between 1650 and 1789, however, little was accomplished. As this chapter will make clear, attempts to develop drainage under the absolutist monarchy were tightly bound

[1] Under Louis XIV, for example, the Crown started to review each project administratively and to evaluate both its legal and physical feasibility. Most projects have left some records in the national archives and an abundant dossier in the regional administration, which is found in the departmental archives.

up with the seigniorial reaction. Thus, for Henri Sée, as well as for Georges Lefebvre, draining marshes was yet another example of growing inequality in Old Regime France between 1760 and 1789.[2] In the case of drainage and irrigation, however, the seigniorial reaction was largely a failure in that improvements were stymied by litigation. Overall, I distinguish four historical periods for investment in water control: the period before 1700, the period between 1700 and 1789, the Revolutionary period itself, and finally the period after 1815. While irrigation and drainage have similar histories, the process of development and the cause of failure were sufficiently different that we should consider them separately.

––––––

Between the fall of the Roman Empire and 1700 there was little development of irrigation except during the thirteenth century.[3] At the end of the twelfth century, a number of religious organizations received water grants from the rulers of southern France. They used the grants to build irrigation canals on both sides of the Durance River in southeastern France.[4] During the medieval population peak, other smaller projects seem to have been undertaken throughout southern France. However, no further irrigation canals were built until the 1760s, with the exception of a canal in lower Provence, which was constructed in the sixteenth century. This canal was an economic success for the region, yet it proved to be the ruin of its promoter.[5]

Unlike irrigation, drainage before 1700 has a complex history. Lay and clerical aristocrats were arguably the most tenacious and successful developers of new agricultural land in premodern Europe.[6] Their participation in these activities reflects both the prejudices that weighed against their direct participation in most sectors of the economy and the fact that they controlled the bulk of undeveloped land and most of the capital in preindustrial Europe. Abbeys played an important role in draining the Pays d'Auge in Normandy as well as a large part of southeastern Provence before 1400.[7] More generally, Benedictine and Cistercian abbeys founded on marshland reclaimed vast amounts of land from the twelfth to the fourteenth century.[8] Lay lords appear to have participated in drainage activities indirectly by granting land to be reclaimed to individuals or to religious organizations. The process of clearing and reclamation was

---

[2] Sée (1906, 208–40); Lefebvre (1959).
[3] For an international perspective on irrigation and drainage, see Ciriacono (1990).
[4] Barral (1876, 370–5); Caillet (1925, xiii–xv).   [5] Bertin and Autier (1904).
[6] Duby (1974, 203–10).
[7] For example, both the Abbey of Troarn in Normandy and the Abbey of Montmajour in Provence were built in the middle of marshland that was subsequently drained under the supervision of the monks.
[8] Dienne (1891, 77–8).

brought to a standstill by the dramatic fall in population during the middle fourteenth century, which drove up wages and reduced the benefits of reclaiming land.

Between 1350 and 1600, local authorities seem to have been little concerned with either improving the quality of already drained fields or extending the cultivated area by draining marshes. Similarly, the royal administration made few direct efforts to improve agricultural productivity through reclamation. This lack of interest in water control is not surprising given the low population levels after the Black Death, the destruction of the Hundred Years' War, and the uncertainties and hardship provoked by the Wars of Religion in the sixteenth century. Thus, one can argue that, between the end of the Roman Empire and the reign of Henri IV, the pace of drainage and irrigation development followed the movement of population.[9] High population levels corresponded to low wages and high land values, both of which encouraged reclamation.

The administration of Henri IV initiated a new phase in the regulation of drainage. By Henri's reign, it was well known that the Dutch possessed the most advanced drainage technology and that a successful transfer of Dutch drainage techniques would allow the reclamation of vast amounts of land in a number of French provinces – in the South near Arles, in the West near Nantes, in Normandy near the Mont St. Michel, and on the northeastern coast.[10] In 1599, to secure Dutch techniques and perhaps Dutch capital, Henri IV granted the newly created office of *maistre des digues* to Humphrey Bradley, a famed Dutch engineer.[11] This grant gave Bradley a monopoly over all drainage projects in France. Bradley was expected to form a drainage company to carry out reclamation projects. The company was given authority to drain any wetland, public or private, for which it would receive half of the drained land.

Bradley undertook a number of projects, mostly in the west of France. Few of these succeeded, however, in the face of staunch local resistance. Indeed, local magnates contested the right of the king to force them to improve their land.[12] Resistance to Bradley's drainage plans was so strong that his grant had to be reaffirmed by the king in 1607, 1611, and 1613.[13] But by 1639 Bradley was dead, and only a few drainage projects had actually been built.

Bradley's death only created further institutional complications for drainage projectors. Title to his 1599 grant was contested, and most of

---

[9] In fact, the only irrigation canal built in Provence between 1350 and 1750 was constructed in the 1550s at the end of the French Renaissance and just before the Wars of Religion broke out. See Chapter 7.

[10] Ciriacono (1989, 99–114).

[11] Poterlet (1817, 1–13). Bradley had worked both in the Netherlands and in England on the great fen project; see Darby (1983, 47–8).

[12] Dienne (1891, chap. 2).   [13] Poterlet (1817, 1–37).

its provisions were substantially weakened by royal intervention, no doubt to quell local opposition. Among the contestants, Noel Champenois claimed to have acquired the grant's authority for a small part of western France including Poitou, Saintonge, and Aunis. He faced competitors in Pierre Siette and Octavius de Strada, who jointly sought to have the king award them exclusive right.[14] In 1640 the Crown eroded the grant's authority further so that any landowner could refuse to participate by promising that he would drain his land himself. Thus, by the middle of the seventeenth century there were at least three claimants to Humphrey Bradley's *privilege* while the value of the *privilege* was itself questionable. These institutional complications, coupled with uncertain feudal claims to land, spelled doom for most drainage projects.

Despite the complications associated with Bradley's grant and local resistance, Bradley's successors did manage to complete a few projects in western France. Projectors were most successful in reclaiming land in provinces where property rights to land were most well defined and most concentrated. In the absence of absolute authority, local opposition could be disarmed only through negotiation. Concentrated ownership and well-defined property rights both reduced the number of participants in the negotiation process and clarified each individual's bargaining position. Marshes in lower Poitou are a good example of successful seventeenth-century operations. Poitevin marshes were usually owned by a very small number of proprietors; thus, bargaining over drainage contracts was relatively simple and speedy. Moreover, these marshes were deep and unused save by fishermen. Fishing rights, unlike grazing rights, were given on a term basis by the owners of marshes. Thus, projectors had to negotiate only with the owners of wetlands and not with users such as villagers.

The case of the marsh of Petit-Poitou provides a perfect illustration of the process.[15] In 1640 the bishop of Maillezais agreed to have his share of the marsh drained by Pierre Siette – the ultimate recipient of Bradley's grant for Poitou. Within a year the Abbey of Moreille, the other major owner of the marsh, signed on as well. Litigation was limited to bargaining with the other magnates of the area: the local chapter of St. John of Malta, the bishop of Poitier, the Abbey of Nieul, and the priory of St. Radegonde. All seem to have agreed to drainage by 1643. Since the marsh was quite deep, it had little value as pasture, and hence there were no use rights to be compensated. Drainage was carried out swiftly, bringing more than six thousand hectares into cultivation.[16] Other Poitevin reclamation projects followed a similar path. Thus, drainage increased

---

[14] Dienne (1891, chap. 1).
[15] Near Fontenay-le-Comte in the contemporary *département* of the Vendée.
[16] Riou (1987, 57–66).

land supply in Poitou by several thousand hectares during the early seventeenth century.[17]

Where property rights were either more dispersed or more uncertain, drainage proceeded with great difficulty or not at all. Such was the case with the Bordelais wetlands, where Bradley's associates struggled to drain several marshes between 1620 and 1650. Although the marshes were not used by peasants and therefore were not subject to customary rights, projectors had to contend with competing claims of feudal ownership. Indeed, the archbishop of Bordeaux laid claim to all the marshes, while local lords each claimed title to the marsh in the village they controlled. Litigation between feudal claimants to land prevented most reclamation. As a result, success in implementing drainage near Bordeaux was much more limited than in Poitou.[18] In fact, it is unclear whether any land was actually reclaimed.

In other parts of France, projectors faced myriad claimants to marshes, and their projects either were never begun or were halted before they had significantly altered the use of the land. Almost everywhere, lords had given villages access rights to marshes, and in many areas a number of parishes had claims over the same marshes. Thus, projectors had to wrestle with two levels of uncertain property rights. As in the Bordelais, a first stratum of conflicts came from frequent seigniorial disputes over ownership of marshes. A second layer of contention concerned the ability of projectors to rewrite customary or use right of villages. Apparently as a result of all the resistance, the Crown interfered more and more against the inheritors of the Bradley grant. Thus, by the late seventeenth century, projectors had no authority to overcome the muddle of property rights and jurisdictions. An example from Auvergne shows just how thorny this situation could become: A follower of Bradley who attempted to drain land near Riom was imprisoned by his opponents and died in jail awaiting royal intervention.[19]

Despite some successes, seventeenth-century drainage projects made only a short-term contribution to reclamation because most projects deteriorated rapidly. The lack of long-term success seems to have been caused by two different forces. The first, subsidence, was physical, while the second, lack of maintenance, was institutional. As the seventeenth-century drainers of the Great Fens in Britain realized, peat marshes sink when they are drained, so that drainage is always temporary and capital improvements are constantly required.[20] Subsidence of this kind may well have been a problem in northeastern France, although the evidence is unclear in this respect. The most important cause of dereliction was probably not physical but lay instead with the absence of an efficient

---

[17] Dienne (1891, chap. 2).   [18] Ibid. (119–38).
[19] Ibid. (chap. 3).   [20] Darby (1983, chap. 4).

organization to assess fees for maintenance. Thus, even routine cleaning of ditches was frequently carried out improperly.[21] In the end, seventeenth-century governments appear to have been unable or unwilling to simplify property rights in agriculture. By adding a new layer of property rights through the creation of Bradley's *privilege*, these governments actually worsened the problem.

The last century of the Old Regime saw growing interest in irrigation and drainage among both landowners and projectors. Nearly all of the proposed irrigation projects aimed to increase water supply in southeastern France, and the vast majority of the new canals were planned specifically for Provence.[22] As we shall see in Chapter 7, all the irrigation canals faced litigation both before and after they were built. Disputes raged both over the apportionment of costs and over the attribution of eminent domain authority. Nonetheless, irrigation was unanimously regarded as a great social improvement.

Nobles promoted more than half of all canals built in the eighteenth century and loaned funds to most other canal projectors.[23] Unlike drainage, where seigniorial intervention was most frequently interpreted by villagers as oppressive, irrigation may have been the most popular of aristocratic activities. Reports of local administrators describe the widespread support enjoyed by these noble projects in terms that curiously anticipate Revolutionary egalitarian ideology:

The lords of this land headed this project [the canal at Aubepagne in southeastern France], loaned money, solicited the government's help and participated directly: the inhabitants today bless these citizen-lords, who have increased their revenue, banished poverty and made the happiness of all the inhabitants [of Aubepagne].[24]

Building irrigation canals appears to have been limited to the southeast between 1700 and 1789. Yet during this period, in almost every French province projectors publicized more ambitious plans of water control. The goal of these plans was to drain and irrigate low-lying areas simultaneously. Yet while the promotion of wetland drainage was a national affair, it was also, as we shall see, a national failure. Moreover, the burst of reclamation proposals had political implications: Because drainage in-

---

[21] Dienne (1891, 81, 130–3).
[22] The exceptions to this rule were three small irrigation canals in the Dauphiné that appear to have been built in the late 1770s. Each of these canals irrigated no more than two or three hundred hectares; AN H[1] 1494. fl. 110, October 8, 1782.
[23] In all fairness to the nobles, it should be noted that they also formed the greater number of those opposed to irrigation development. See Chapter 7 and AN H[1] 1494, fl. 110, 133.
[24] AN H[1] 1494, fl. 133 (August 1781).

volved a change in the ownership of land, it was frequently seen as an abuse of power by projectors and feudal lords.

Political decisions were at the heart of both the eighteenth century's renewed interest in drainage and the failure of most projects. During the late seventeenth century, the Crown's increasing regulation of common land made it more difficult to drain marshes. In the early modern period, state expenditures rose rapidly, and so did taxation. At the same time, many heavily taxed peasants sold their land to tax-exempt nobles and city dwellers.[25] The land-sale movement of the early modern period dramatically eroded the royal tax base and thus increased rural tax rates. Since taxes were assessed not only on private holdings but also on common land, it was in the interest of villagers to sell their common land so as to decrease their collective tax burden. Villages also found it necessary to sell their common land to pay debts incurred in times of war. Tax-exempt lords, who were frequently the largest creditors of villages, bought most of this common land. Late in the seventeenth century, the Crown realized that it was necessary to put a stop to the vast land transfer because it was eroding the tax base. In 1656 and again in 1677, legislation was enacted to stop the sale of common lands. These new laws made future sales of common lands and common rights contingent on royal assent. The reforms also allowed villages to buy back commons sold after 1556, at their original price. In practice, however, it appears that little of the common property sold during the sixteenth and seventeenth centuries was recovered.[26]

Since wetlands were not very valuable, they were relatively unaffected by the vast movement of land sales. Thus, most often villages retained their use rights to marshes. Wetlands were, however, dramatically affected by the laws aimed at protecting the royal tax base. These laws made it virtually impossible to alter the state of common property.

By the 1660s most marshes were, in appearance at least, common land. Over the centuries, lords who judged drainage unprofitable had allowed villagers to use shallow marshes for pasture and other agricultural activities. Seigniors had often allowed villagers access to land that was not common. Such land was usually of little value and called waste. Both common land and privately held wasteland were used primarily for logging and pasture, and as a result the two could be easily confused.[27] Because of this confusion, the reforms of the 1660s made it difficult for

---

[25] Hoffman (1986).

[26] Isambert, Jourdan, and Decrusy (1821–33, Vol. 17, 372). The effectiveness of the law seems to have been mixed. According to Saint-Jacob, seigniors in Burgundy continued to "usurp" common rights until the end of the Old Regime. In contrast, Lefebvre argues that, in the Nord until the 1770s, there was little change in common lands. Cf. Lefebvre (1959, 100–1) and Saint-Jacob (1960, 144–7).

[27] See, e.g., Sion (1909, 202–19).

lords to recover wasteland. After the reform of 1677, seigniors could recover wasteland only if they submitted to a long, complex, and costly judicial determination of ownership. Villages, in turn, could not sell their common land to a drainage projector without royal assent. Thus, most undrained marshland was frozen in an unimproved state.

The edict of 1677 had not made clear what rules would decide whether land would be deemed waste or common. An effort to clarify the ownership of common land and wasteland was made in 1699. The Crown used its review of the offices of the Eaux et Forêts to set down new rules for *triage*, the rule that governed the division of pasture- and forestland between lord and village. Title 25 of the *Ordonance des Eaux et Forêts* set the conditions under which woods, marshes, and pastures could be divided:

Art. 4: If the woods [of a parish or community] have been freely conceded by the seignior, without charge of *cens,* dues, prestations or servitudes, a third may be subtracted from the woods and separated for [the seignior's] benefit if he requires it and if the two-thirds are sufficient for the need of the parish. . . . This will be equally observed for grassland, marshes, islands, moors, and pastures. If the lands were not freely conceded, then the seignior will only have use rights and the right to pasture his animals as first inhabitant, without partition or *triage,* nor prestations, dues, or servitudes.

Art. 5: Concession will not be deemed free on the part of the seignior if the inhabitants can justify that they have acquired [the land] even if they are held to no dues. If they had paid some [rent] in money, service, or in any other fashion, the concession will be deemed costly, even if the inhabitants cannot find title to the land, and this will prevent all subtraction to the benefit of the seignior who will only receive his use rights and firewood as is customary.[28]

This legislation proved to be the Old Regime's definitive statement on the process of *triage*. The 1699 rules implied that lords could initiate *triage* — and thus drainage — only if they could prove that they had, through the centuries, retained complete or ultimate property rights to marshes. Such a demand was almost impossible to satisfy. In only two types of cases did lords command sufficient evidence to order *triage*. First, deep marshes were most often let on term leases and were thus never confused with the commons. Second, a few religious institutions managed to preserve such detailed records dating back to the Middle Ages that they could show in court what land was common and what was waste.

From 1699 to 1740, lords and other promoters of drainage seem to have received little or no backing from the royal government. Drainage projects were almost invariably initiated by lords in order simultaneously to achieve *triage* and to increase the value of their holdings by improving

---

[28] Isambert, Jourdan, and Decusy (1821–3, Vol. 17, 181–2).

water control. Reclamation projects were held up by endless litigation as courts in Paris and elsewhere attempted to resolve ownership disputes judicially.[29] The records of these disputes indicate that no one doubted the profitability of drainage; the source of contention was the allocation of the benefits from improvement. While the royal administration participated in these disputes, it seems to have been incapable of resolving them.

During the 1760s, the royal administration's attitude began to change. The government became more favorable to property rights reforms that would increase agricultural output. Reforming ministers argued that land should be held privately rather than communally because they felt individuals would take better care of private fields.[30] *Triage* and drainage were methods of attacking common property rights that had the additional benefit of considerably raising the productivity of the land. Royal reformers first attempted to solve the legal problems of drainage projectors with a set of national decrees. This legislative activity was part of an ambitious plan to increase agricultural output that also included tax rebates for clearings and attempts to auction off royal lands to potential developers.[31] The drainage proclamation of 1764 promised royal support for reclamation projects, sought to limit to six months the time during which inhabitants could object to an improvement proposal, and granted large tax exemptions to would-be drainers.[32]

The proclamation, however, offered no relief from the fundamental problem that stymied drainage projectors – litigation. In many provinces, the legislation on marshes fared little better than that on enclosure, which as Philip Hoffman put it, "was emasculated" by *parlements* and other local organizations.[33] Faced with the failure of their national reforms, royal administrators tried to promote change in those provinces where common lands were most prevalent. Thus, legislation requiring *triage* and separate edicts promoting drainage were enacted in the 1770s for a number of regions. In addition, the administration carefully monitored the scores of projects that sought special legislation because even regional edicts were insufficient to stifle litigation.[34]

In the late eighteenth century, the combination of new legislation and rising agricultural prices led to a host of projects, most of which failed because of litigation. Except for the Southwest, every area of France seems

---

[29] For yet another example of failure through litigation in central France, see Poitrineau (1979, 325–6). For a few examples of success coming after decades of litigation, see Lefebvre (1959, 79).

[30] Bloch (1929).

[31] Ironically, the same uncertainties that plagued seigniorial property also plagued royal lands, and the auctions led to little drainage and lots of litigation. AD Calvados C 4277.

[32] AN H¹ 1499. fl. 68.   [33] Hoffman (1988, 241–64).

[34] Isambert, Jourdan, and Decrusy (1821–33, Vol. 27, 233–4).

to have been engulfed in a drainage boom. Table 4.1 displays the fate and size of those projects that I have been able to track down for each major French province. As the table makes clear, only a handful of projects succeeded. Those that did were of three kinds. Some drainage was financed directly by the royal government, primarily to reinforce the defenses of the eastern border or to reduce the unhealthiness of some ports on the western coast. These projects were few, and it has been argued that many were mismanaged.[35] Some drainage projects were undertaken on completely submersed lands, mostly in Poitou and Flanders. These projects, financed by local organizations or by individuals, seem to have been somewhat successful. The vast majority of projects involved common and shallow marshes. These reclamation efforts were undertaken by private projectors. In such cases, litigation was prolonged, bitter, and expensive, and successful drainage was the exception rather than the rule.

Drainage was a highly divisive issue both at the local level and within the royal administrations. On one side of the reform issue stood the judiciary, frequently supported by tax collectors. The judiciary opposed drainage because it threatened local *privileges* and made tax collection more difficult. The judiciary had political motivations for opposing drainage as well: Royal policy in agriculture was part of a more general attempt to increase uniformity in France through the use of national legislation.[36] Since judicial officers gained much of their prestige and wealth from their role as mediators between the state and the provinces, any attempt to bypass them was met with deep suspicion.[37]

On the other side of the reform issue, central administrators often championed agricultural innovation. They produced reports advocating drainage and faulting the nation as a whole for the failure of reclamation efforts:

To cultivate those lands which remain unused in our inheritances is to acquire real assets, and to increase one's possession. Useless land is saddening, even dishonorable for the nation since it can only be the result of the laziness of the poor who prefer shameful begging and the rotten life that they lead to the honest well being which work brings. [Uselessness may alternatively be due] to the carelessness of landowners, which may be even less excusable since it would appear that they are indifferent spectators to the idleness of the poor. They would see them suffer instead of giving them work.[38]

Within villages, drainage proposals were highly divisive because the rules for allocating property left little room for negotiation. In effect, there were only two claimants to marshes: lords and villagers. The legis-

---

[35] Lefebvre (1959, 226).
[36] Other examples include the Maupeou reforms of the judiciary and the municipal reforms of the 1780s; see Jones (1988, 27–9).
[37] Mousnier (1979, 1:609).
[38] AN H[1] 1495 (from a report on a project to drain marshes in Artois, 1780s).

Table 4.1. *A survey of eighteenth-century drainage projects*

Province	Marsh	Project Start	End	Outcome	Area[a] (hectares)
Artois	Gondecourt		1773	Division, no drainage	
	Moere	1640	1790	Drainage never completed	
Aunis	La Boutonne	1766	1781	Contract never executed	1,100
	Lande	1786	1789	Contract never executed	
Auvergne	Riom	1719	1768	Litigation, no drainage	
	Riom	1782		Litigation, no drainage	
Bresse	Eschelles	1784		Litigation, no drainage	1,151
Brittany (North)				No attempts at drainage Similar problems with moors	
(South)	Donges	1774	1780	Litigation, no drainage	1,239
	Grand lieu	1714		No drainage until after 1799	2,800
Burgundy (North)				Marshes are rare Similar problems with woods	
Dauphiné	Bourgoin	1766	1789	Litigation, no drainage	5,000
	Luc	1752	1789	Litigation, no drainage	
Flanders and Hainaut	Trith		1733	Divided and drained	2,000
	Onze-Villes	1759		Divided, no data on drainage	
	St. Amand	1764	1782	Drained by a royal project	71
	Dechy	1777		No drainage, lord gets payment instead of division	
	Santes	1760			
	Maing	1762			
	Arleux	1778			
	Hasnon	1761	1772	Litigation, no drainage	
	Lescluse	1770	1789	Litigation, no drainage	
	Sainghin	?	1789	Litigation, no drainage	2,000
	Scarpe Marque	1779	1789	Tax exemption for drained fields; some drainage	2,600
Guyenne and Gascony				Litigation, no drainage	
Languedoc	Marseillac	1760	1789	Litigation, no drainage	
Limousin				No common land, no drainage	
Normandy (East)	Vernier	1761	1772	Litigation, no drainage	2,500
Normandy (West)	Barneville St. Samson	1712	1773	No drainage, no division	200
	Ecramville	1780	1789	Drainage not completed	
	Boucey Songeal	1774		Litigation, no drainage	

Table 4.1. *(cont.)*

Province	Marsh	Project Start	End	Outcome	Area[a] (hectares)
Picardie	Authie Voisin	1782	1787	Failed because of disagreements	1,000
		1785		Drained after 1787?	
Poitou	St. Morick	1728	1730	Litigation, drainage	4,393
	Brierre	1770	1784	Litigation, no drainage	7,293
	La Claye	1704	1742	Litigation, no drainage	1,941
	La Crosnière	1767	1772	Drained	250
	Ollone	1774		Litigation, no drainage	300
	Pironnerie	1778		Drained?	79
Provence	Berre	1779	1783	Litigation, no drainage	201
	Marignane	1750s		Drained	300
	Arles	1770s	1789	Litigation, no drainage	6,000

[a] One hectare equals 2.4 acres.

Sources: For Artois, Lefebvre (1959, 79–80, 225–7); Aunis, AN H^1 1496 (La Boutonne) and AN H^1 1497 (Landes); Auvergne, Poitrineau (1979, 324–6); Bresse, AN H^1 1497; Brittany (North), Sée (1906, 208–40); Brittany (South), AN F 10 319; Burgundy, Saint-Jacob (1960, 44–147); Dauphiné, AN F 10 208; Flanders and Hainaut, Lefebvre (1959, 79–80, 225–7), and Fruit (1963, 53–6); Guyenne and Gascony, Dienne (1891, 171–221); Languedoc, F 10 319; Limousin, Laffarge (1902, 50–1); Normandy (East), Sion (1909, 202–19); Normandy (West), AN H^1 1492, AC Calvados, C 4226–2450; Picardie, Poterlet (1817, 245); Poitou, Riou (1987); Provence, Masson (1929, Vol. 7).

lation that governed division allowed only three kinds of allocation: all to lords, one-third for lords and two-thirds for villages, or all to villages. When marshes were large, changing the allocation could lead to dramatic differences in the value of drainage for each party.[39] Thus, villagers and lords alike were willing to fight in an attempt to increase their share of a marsh. Moreover, division and drainage would have changed the way a marsh was classified: common or private. As noted earlier, such a change had the potential to affect many villagers.[40]

An important factor behind the obstructionism of local groups was the state's constant willingness to renegotiate. In different regions, different

[39] In Chapter 9, I will return to these division rules to analyze why they increased litigation and prevented drainage.

[40] Jean Meuvret (1987, 2:195–7) suggests that large landowners, rather than small landowners, often received the lion's share of common rights. See also Saint-Jacob (1960, 369–84).

groups were able to have legislation or drainage contracts significantly modified as a result of their opposition. For example, in Artois during the 1770s new *triage* edicts that gave lords one-third of all communal marshes were stridently opposed by villages. This opposition had significant effects – in the 1780s the division rules were revised again, and unimproved marshes were again subject to the 1669 ordinance. Henceforth, lords would receive marshland only if they could prove that marshes were part of the waste. In all other cases, there would be no *triage*.[41]

In Artois, it was villages that resisted reforms. In Burgundy, lords were the ones who opposed new legislation. They appear to have enjoyed more than a third of the common and as a result were large owners of livestock. To protect their grazing rights, they opposed any limitation to their access to the two prime sources of pasture: the fallow and common land. As a result, proposals to divide common land in Burgundy were shelved indefinitely.[42] In the absence of a rule for separating common and private land, little drainage occurred. Finally, in the Norman village of Ecramville, smallholders opposed drainage rules enacted in 1787 that specified that the village's portion of the marsh would be divided according to landownership. As a result of their opposition, new rules more favorable to smallholders were enacted.[43] These new rules were attacked roundly by large landowners and, not surprisingly, went down to defeat as well. Clearly, in the eighteenth century, irrigation and drainage were profitable yet contentious. What was lacking was a mechanism to stop litigation and allow improvements.

———

Little investment in irrigation or drainage occurred between 1789 and 1815. Very few marshes seem to have been drained during this twenty-five-year span, and no irrigation canals were built.[44] One can easily attribute the absence of activity in water control in the decade after 1789 to the turmoil of the Revolution. The lack of investment during the next fifteen years (1800–15) was probably the result of unfavorable economic conditions and institutional instability. Although Napoleon sought ceaselessly to legitimize his regime and solidify the property rights structure created during the Revolution, investors may have been unconvinced. They probably withheld their confidence from the new regime of

---

[41] Sueur (1982, 2:713–15).   [42] Saint-Jacob (1960, 378–80).
[43] AN H¹ 1489, fl. 88.
[44] In 1808, the Ministry of Interior was reorganized, and authority over drainage projects was transferred from the Department of Agriculture to a new Department of Drainage. Nearly all the projects in progress were ones that remained from the Old Regime (AN F 14 6391). The abandonment of drainage and irrigation reflected a more general disaffection with water control. Dion (1961, 205–21).

property rights until the Bourbon monarchs returned and agreed to the Revolutionary changes, including the destruction of feudal property in 1815.

Development of water control was in fact contingent on the elimination of feudal property rights. As already noted, before the Revolution lords had had extensive but disputed claims to the vast bulk of marshes in France: common marshes. Lords similarly held disputed title to water rights and eminent domain authority. Both of these property rights were crucial to the development of irrigation. The Revolution permanently clarified authority and property over land and water. Feudal claims to marshland were transferred to villages, while claims to water and rights of way were preempted by the state.[45] Moreover, villagers were given the right to divide their commons if a majority of the heads of households agreed to it.[46] Hence, not only were property rights reordered to decrease conflict, but a simple rule for dividing common property was enacted, a rule that held for all of France. This rule was not negotiable on a case-by-case basis. In short, nearly all causes of contention had been removed in one fell swoop.

The new distribution of property rights, however, would lead to more irrigation and drainage only if investors could be convinced that it would be upheld by the state. The problem of security was twofold. Not only were Revolutionary regimes themselves fragile and unstable, but they also acted in ways that, at first, increased uncertainty in property rights. Indeed, Revolutionary governments were often in dire political straits, and they sought to use their power to increase the chances that their regime would survive. They intervened in the economy to regulate markets because of political necessities, and that increased uncertainty in property rights.

A pertinent example of Revolutionary regulatory activity is the pond law of 28 Brumaire of the year II. Having faced scarce grain supplies and high grain prices from 1788 through 1792, the Convention (the national assembly of 1792) tried through legislative means to increase agricultural output. To this end, the Convention decreed that all ponds should be drained and sown in *legumes* (peas and beans) or in spring grains. Only very small ponds and those that served defensive or manufacturing purposes were exempted from this legislation.[47]

The pond law affected northern France primarily because in the South very few ponds existed. Compliance with the law, however, would have

---

[45] Merlin (1828, 6:397–451, 16:105–46).

[46] While rules that did not make villagers pay for their share of divided common land were in effect only briefly, other reforms did not stand in the way of the division of common land because they required only that each individual buy his share at a low price.

[47] Gerbaux and Schmidt (1908, Vol. 3, 153–4).

threatened an important sector of the northern rural economy: fish farming. Most ponds were alternatively "wet" and "dry." Ponds with water were stocked with fish. After a few years, they were drained and sown with small grains for one or two harvests. The wet and dry phases of the rotation were complementary: Land was fertilized during the years when ponds had water, and each time a pond was filled the residues from grain cultivation fed the fish. The pond law of year II interfered with this system by forcing pond owners to dry up their ponds to increase grain production. Most owners searched for ways to avoid the law. To control and increase compliance, the Convention sent a number of inspectors during year II and year III to the provinces. Their reports suggested that the law was misguided because most ponds were vital to the local economy as sources of water power or as sources of water for animals.[48]

In its aim, the law on ponds was a direct echo of the national agricultural edicts of the 1760s. Yet in its effectiveness, the law of the year II was revolutionary. Old Regime legislation had been subject to complex bargaining between the state and regional authorities and was thus at risk of local amendment or abandonment. Laws enacted by the Convention faced no such constraints. Although the law on ponds was eventually annulled once the complex role of ponds in agriculture became clear, it was for a time applied all over France. Similarly, all complaints were handled administratively, not judicially, and according to national, not regional, standards. The law on ponds illustrates both the benefits of the Revolution and its risks. The Revolution made possible speedy and uniform institutional change, but the newfound power of the state carried the risk that legislators would intervene in the economy in ways that would slow rather than speed development.

The Napoleonic regime only increased the effectiveness of the administrative structure and the authority of the central government beyond what had been achieved in the first few years of the Revolution. The fundamental law governing drainage was passed in 1807. It was to become the covering law for all land improvement projects until the 1860s.[49] The law gave simple rules for drainage and empowered the government to force reclamation if landowners could not agree to carry out projects by themselves. More important, it allowed projectors and village councils to contract freely over most aspects of drainage. Once a contract had been reached, the state's local agent, the *prefet,* would enforce it. The only requirement placed on improvement proposals was that they create an organization to carry out maintenance and further improvements. This organization, called an *association* or a *syndicat,* would be supervised by

---

[48] AN F 10 308–318. Only in the Orléanais does it appear that the law increased the number of ponds under cultivation.
[49] See Poterlet (1817).

the state. Negotiations, contracting, and the resolution of disputes were all carried out under the authority of the *prefet* and his technical aide, the engineer of the Ponts et Chaussées.

The period from 1815 to 1860 was one of modest institutional change in water control; nonetheless, many marshes were drained and many irrigation canals were dug. Let us first consider the evolution of property rights after 1815. The 1807 legislation on drainage remained in force throughout the nineteenth century. Overall, laws concerning the division of common land and agricultural associations underwent only minor revisions in the 1820s. The primary institutional effort toward increasing water control was administrative rather than legislative. For example, administrators extended the 1807 law to cover a wide variety of agricultural improvements. As a result, the *associations* of the law of 1807 that were in fact Old Regime innovations became widely used in agriculture after 1815. Before the Revolution, each *association* had independent statutes. The degree of administrative protection and oversight also varied greatly from one project to another. One source of contention under the Old Regime was the choice of rules that would govern the *associations* overseeing drainage and maintenance. After the Revolution, both old and new *associations* received a similar degree of protection and supervision from the state. In effect, the laws governing *associations* were used by administrators to systematize the relationship between landowners, projectors, and the government.

After 1815, development proceeded on a number of fronts. While landowners grouped in *associations* sponsored most of the new small-scale water control activity, private corporations were created to improve large tracts of land. These private firms carried out large-scale reclamation in areas where marshes were deep and in some cases attempted to build irrigation canals. Unlike Old Regime private arrangements, these companies were national and frequently operated a number of projects at the same time. Another source of drainage activity was the state itself. In the 1807 law, the state had reserved the right to intervene in local affairs and mandate reclamation. This right was exercised under three different circumstances. The state intervened when private development failed, when reclamation would substantially reduce water stagnation, or in order to improve the transportation network.[50] The state intervened most visibly in the South because marshes there were often breeding grounds for malaria.[51] In the case of common marshes, which were gen-

---

[50] Dion (1961, 219–24); Sion (1909, 338–42).
[51] Masson (1929–30, Vol. 7, chap. 10).

erally small and posed no health threats, private solutions for reclamation prevailed.

In the nineteenth century, evidence on irrigation and drainage began
to be collected systematically. Thus, we can assess the magnitude of nineteenth-century accomplishments in water control by a quantitative analysis of changing land use. Most measures of reclamation efforts in the
years between 1815 and 1860 suggest that, relative to the total amount
of land in France, the growth in agricultural land supply was limited.
This negative conclusion, however, is misleading. First, it does not hold
up regionally, since some areas experienced substantial acreage gains
through increased water control. Second, the relatively small size of the
gain should not be taken as an indictment of the Revolution's institutional reforms, because these reforms affected many sectors in the economy other than marginal land. In fact, since France had a diversified
economy, it would be unlikely that any one institutional change would
increase national output by more than a limited amount. Finally, one
should bear in mind that the appropriate comparison for the period from
1815 to 1860 is the final seventy years of the Old Regime, when nearly
nothing was accomplished.

Let us first consider drainage. On the basis of cadastral and census
survey evidence, Hugh Clout estimates that wasteland declined from 8.5
million hectares in 1837 to 5.9 million by 1862, or by nearly 30 percent.[52] The gain represented 8 percent of all arable, vine, and grassland
in France. Since marshes were not the only source of wasteland, drainage
could only have been smaller than the whole of wasteland improvement.
Yet the national totals understate the reclamation effort because the acreage
of poor land that was abandoned between 1837 and 1862 counted against
the acreage of better land that was brought into production. Presumably,
reclaimed land was more productive than land that was abandoned, and
as I shall argue later it was likely more productive than average. Thus,
replacing low-quality land with reclaimed land would raise the average
productivity of land. Such an increase in average productivity would be
masked by aggregate statistics that provide data only on net changes in
land use. As a result, wasteland reclamation may well have contributed
more than 8 percent to the growth of output in the early nineteenth century. Nevertheless, national statistics on wasteland reclamation are too
aggregated to determine the contribution of water control to national
output.

Since marshes were not the dominant source of reclaimed wasteland,
a more direct estimate of marsh drainage is required to appreciate the
contribution of the Revolution's reforms. An 1817 survey estimated that

---

[52] Clout (1983, 46–9).

425,000 hectares of land still needed to be drained.[53] Regional checks suggest that most of the marshes included in the survey were drained before 1870. In fact, 425,000 hectares is probably a severe underestimate of undeveloped French marshland for the early nineteenth century. The 1817 figure, based on a *département*-by-*département* survey, is inferior by as much as 50 percent to estimates of nineteenth-century reclamation efforts that appear in local studies. For example, in the Bouches du Rhône, the survey reported 53,000 hectares of marshland. The *Encyclopédie des Bouches du Rhône,* however, published in the 1930s, revealed that more than 70,000 hectares of wetlands were reclaimed between 1815 and 1870.[54] Similarly, in the Calvados the 1817 survey listed 3,240 hectares of marshland, yet other sources suggest that there were more than 4,000 hectares of deep marshland, and that many shallow marshes were classified as moors; as a result, Calvados marshland may have been underestimated by 50 percent.[55] More important, while much of the deep marshes remained undrained until after 1870, most of the shallow ones were improved by 1860. In the *département* of the Eure, the survey counted 2,500 hectares of marshes when some 6,700 hectares were brought into cultivation between 1814 and 1848.[56] If we adopt the larger figure rather than that of the survey, reclamation in the *département* of the Eure would nearly triple, raising the increase in arable land there from 1 to 2.6 percent. In each of these *départements,* most of the marshes listed in the 1817 survey had been drained by 1860. On the whole the 1817 survey can be used as a conservative estimate of nineteenth-century drainage effort.

Although the 1817 survey underestimates drainage achievements in early nineteenth-century France, the survey's probably very low estimate of 425,000 hectares of marshland represented nearly 1.5 percent of all French arable and pastureland.[57] A perhaps more relevant comparison involves pastureland, since reclaimed marshes were most often sown with grass. Lowlands drained between 1815 and 1860 amounted to 10 percent of all French pastureland. More important, wetland reclamation was equivalent to 50 percent of the increase in pastureland between 1837 and 1862.[58] Lumping all pastureland together probably understates the importance of drained land, which was mostly converted in *pré* – meadowland – the most prized of all agricultural land in France. Thus, drain-

---

[53] Poterlet (1817, 285–315).   [54] Masson (1929–30, Vol. 4, chaps. 9 and 10).

[55] Desert (1977, 310–15).

[56] Vidalenc (1952, 401–16). In fact, the area susceptible to improvement according to the 1814 survey, which in other respects is very reliable, was 25,000 hectares, or ten times what was mentioned by Poterlet's 1817 survey; to be sure, most of this land was not marshy, but probably no less than a fifth was in need of drainage.

[57] It seems fitting to exclude vines, since marshland was never suitable for growing grapes.

[58] Clout (1983, 49).

ing land was a valuable, if not crucial, way to increase agricultural output in the late eighteenth and early nineteenth centuries.

One should also note that reclamation had dramatic regional effects. Marshland was distributed unevenly so that, for at least ten *départements*, reclamation allowed the extension of arable land from 5 percent to as much as 30 percent. If an increase of 2 percent of arable land due to drainage can serve as a lower bound of the impact of drained land on acreage as a result of Revolutionary reforms, 4 percent is likely to be a good upper bound of the impact of the Revolution on agricultural output. Indeed, most drained marshes were easy to irrigate. Where it was possible to control water flow completely, land was highly productive. The value of drained land was especially high in the South because it is difficult to produce goods that demand large quantities of water in an arid area. Thus, in southern France irrigable land commanded twice the price of regular arable land. Substantial premiums were also paid in the North for marshland, which most often could be turned into high-quality pasture. Thus, although not all drained land was twice as valuable as average land, the output gains from drainage are likely to have been more than 2 percent.

Irrigation seems to have had much more limited development in the nineteenth century. By 1863, there were 200,000 hectares of irrigated land in France. Irrigation could have only a limited impact because most of the land irrigated by 1860 had been irrigated before the Revolution. Moreover, nineteenth-century increases in irrigation were limited because of geographic constraints. Most of the irrigated acreage in France was concentrated in mountainous areas where expanding the networks was difficult.[59] In the one area that showed any promise for increased irrigation – the Southeast – water became scarce as a result of the early-nineteenth-century expansion. In all, the increases in irrigated acreage is unlikely to have been more than 25,000 hectares, with two-thirds of this figure coming from Provence. Since 25,000 hectares was less than one-tenth of 1 percent of the total amount of land in nineteenth-century France, it is unlikely that increased irrigation between 1815 and 1860 contributed much to the growth of total output. Yet as Chapter 7 will make clear, all the projects that faltered or failed during the last century of the Old Regime were realized, and these new irrigation canals were of great importance to local economies. For irrigation projects, natural constraints replaced institutional barriers to development, and that alone made the Revolution a success in water control.

The first half of the nineteenth century saw a dramatic increase in both drainage and irrigation activity. This increase did not depend on new technology, as we shall see shortly. In fact, projects were realized with

[59] Barral (1875–6, Vol. 1, 86).

the crudest of methods. Moreover, one cannot argue that the Revolution bred a new type of entrepreneur that was willing to bear the risks involved in agricultural development. Indeed, the individuals who belonged to the *associations* that accomplished most of the nineteenth century's water control projects were most often direct descendants of the individuals who had most strongly opposed the Old Regime projects: local aristocrats, landowners, and farmers. The entrepreneurs who headed drainage and irrigation companies were frequently nobles.

The nineteenth-century experience in water control presents a stark contrast to that of the eighteenth century. It is clear that water control under the Old Regime did not fail because drainage was not attempted, but rather because most proposals drowned in a sea of disputes and litigation. Moreover, the state's attempt to remedy the situation before 1789 usually worsened, rather than alleviated, the property rights problem. There are many possible explanations for Old Regime failures and for post-Revolutionary successes in drainage and irrigation, including relative price changes, technical advances, and institutional reforms. The next three chapters take up the task of discriminating among these hypotheses.

5

Relative prices and the supply of water control

There are potentially two broad reasons for the dramatic success of water control, at least at the local level, after the Revolution. On the one hand, demand for improved land may have increased because land may have become scarce after 1815. On the other hand, supply conditions may also have changed because the costs of improvements may have dramatically fallen during the Revolution. Both could have been equally due to changes in technology, institutions, or relative prices. In this chapter, I begin the task of discriminating between these hypotheses by constructing a framework to evaluate the returns to improvement and by examining relative price data.

I shall begin by examining the supply of improved land. The lack of water control development in the eighteenth century might be explained by institutional or technological factors that could have affected supply decisions. For example, as suggested earlier, during the eighteenth century, institutions might have raised the cost of development enough to discourage investment. Another explanation relies on technical change. Perhaps technological constraints hindered water control before the Revolution, while breakthroughs in the period between 1789 and 1820 could explain the sudden success of improvement projects in the nineteenth century.

The demand for improved land could also have changed significantly over time. French economic historians have long investigated the relationship between population and prices. To a large extent they have concluded that, until the middle of the eighteenth century at least, French society was caught in a Malthusian system: Wages were low and goods were dear in periods of high population. Increased population thus corresponded to periods of relative scarcity of food and therefore of land.[1] Since water control was a method for increasing the supply of land, one could reasonably suggest that the rhythm of water control improvement

[1] Le Roy Ladurie (1966, Vol. 1, pt. 2 and conclusion); Goubert (1960, 1:599–618).

in France responded to the rhythm of prices. Superficially at least, the long-run record of drainage and irrigation is consistent with such a hypothesis, since development coincided with medieval and Renaissance population peaks. Yet the eighteenth century offers a puzzle because, while population was growing and land prices were rising, little improvement actually occurred before 1820.

Holding institutions and technology fixed allows us to carry out an initial investigation of changes in the demand for and supply of improved land. A few simplifying assumptions will enable us to construct a model of water control investment and examine how relative prices might have affected drainage and irrigation.[2] The model is described in detail here because it is the basis of most of the quantitative analysis in this book.[3]

The model is essentially static. Assume that the global production function for improved land has only two arguments – unimproved land and labor – and that it is strictly concave. In other words, we assume that increasing the amount of unimproved land or labor employed in the reclamation process will lead to more improved land but at a decreasing rate – in effect, each extra acre of improved land is more expensive than the previous one. In defense of these assumptions, one should note, first, that the principal costs of irrigation or drainage are labor and the purchase of unimproved land. Second, not all land is equally easy to drain; thus as more land is improved, marginal projects become more costly – the concavity of the production function captures the increasing marginal cost of reclamation.

We can denote the production function of improved land $H(m, L)$, where m is unimproved land and L is labor. Let the price of arable land be p, the opportunity cost of unimproved land p_m, and the wage w. This allows us to write the profit function of drainage as

$$\Pi(m, L) = pH(m, L) - wL - p_m m.$$

Let us denote the supply of improved land by n. Because ditches and canals must be built, the usable acreage of an improvement $(n = H(m, L))$ tends to be less than the unimproved land (m).[4] If we assume that the

[2] Specifically assume that, in all markets, all goods and services (except land, of course) were in perfectly elastic supply and that technology is completely static over the period of investigation. These assumptions will reinforce the relationship between water control and relative prices.

[3] The models are intended to be illustrative. Most of the specific assumptions can be discarded, but doing so would unnecessarily complicate the analysis.

[4] There were losses of land where irrigation and drainage canals were built. Moreover, sometimes parts of the center of the marsh proved particularly difficult to drain and were thus not improved.

Relative prices and the supply of water control

supply of water control entrepreneurs is competitive, then the quantity of land improved, which we can denote by n^*, will maximize the social returns to reclamation.[5] Moreover, simple comparative statics will yield

$$\frac{\partial n^*}{\partial p} > 0 \qquad \text{and} \qquad \frac{\partial n^*}{\partial w} < 0.$$

Thus, in a competitive market the quantity of land improved will depend on the price of arable land (p) and the price of labor (w). An increase in the price of arable land should lead to an increase in water control, while an increase in the price of labor (wages) should lead to less water control.

This model, of course, treats improvements as instantaneous processes, when in fact they are investments. In other words, the model has ignored the cost of one input – capital. Moreover, both irrigation and drainage are discrete goods because each drainage proposal or irrigation project will affect a specific number of hectares of land with a specific improvement technology. The quantity of land improved will therefore depend on rates of return, provided that projectors carry out improvements for profit. Clearly, a better model would explicitly assess the impact of capital costs and differentiate between different projects. To analyze this more complex problem, let us index all projects under consideration. Assume that a fixed quantity of labor L_i is required to complete project i, which, when accomplished, yields n_i units of land and has opportunity cost m_i.[6] Assume that it takes T years to complete the project and that the labor invested annually in the project is $L = T/L_i$. When the improvement is completed, the land is sold at a price of P_i. The price of improved land (P_i) equals the price of arable land (p) minus the discounted present value of maintenance costs per unit of surface. Then the internal rate of return is δ_i such that

$$\frac{n_i P_i}{(1 + \delta_i)^T} - m_i - \sum_{t=0}^{T-1} \frac{l w_i}{(1 + \delta_i)^t} = 0.$$

In this model, all projects with an internal rate of return higher than the interest rate will be carried out. To look at the impact of changes in relative prices, let us order the projects by their internal rates of return, for a given set of prices. It is easy to check that the internal rate of return rises as the price of arable land rises and falls as the price of labor increases. Thus, if the price of land rises sufficiently, a project whose rate of return was initially less than the rate of interest will become profitable.

---

[5] The assumption of a competitive market for improvement supply is warranted by the simplicity of the improvement technology.
[6] That is, marsh i would command a price of m_i in a competitive market.

Drainage and irrigation

A rise in the interest rate has three effects on the supply of improvement: a higher rate of return for a project to be profitable, a decrease in the price of land, and a decrease in the opportunity cost of the project. The second effect must be larger than the third because improved land carries a higher price than unimproved land. Therefore, the net effect is to make all projects less attractive.[7]

Both the simple model and the more complex one highlight the importance of the price of land relative to labor. Thus, a good index of the returns to improvement seems to be an analysis of changes in the relative cost of arable land and labor. Indeed, a dramatic change in the relative prices of land and labor around 1789 would go a long way toward explaining the abundance of projects that were carried out in the nineteenth century.

Collecting information on the price of land is a long and tedious task; as a result, land prices series are rare. However, other data, such as the price of wheat, are readily available because many historians concerned with Malthusian crises have used the ratio of wages to wheat prices as a crude proxy for standards of living. Indeed, in preindustrial France bread was the primary staple, and food was the most important and variable item in household budgets.[8] Similarly, land supply was an important constraint on the growth of grain output. As the price of wheat rose, the profitability of creating more land must have increased. Thus, we can take the ratio of wheat prices to wages as a rough indicator of changes in the profitability of water control.[9] Such a simple approach to the issue of relative prices is advantageous in that it offers a picture of prices on a national level. This is true because early regional integration of grain markets over large subsections of France caused price trends to converge. Thus, a few price series can offer a good long-run picture of wheat prices all over France.

While wheat prices are relatively easy to come by, wage data are scarcer. Published long-run series of wage data for unskilled or skilled labor are few except for Paris.[10] One important reason for the paucity of wage data is archival. Indeed, while as early as the sixteenth century local officials closely monitored the price of grain in many areas of France, no such attention was paid to wages, Hence, there exists no concentrated source for wage rates. As a result, many of the data pre-

[7] Of course, changes in the interest rate may alter which projects are most profitable.
[8] See the now classic volumes of Labrousse (1933), Goubert (1960), and Le Roy Ladurie (1966).
[9] This was the approach advocated in another context in Baehrel (1962, Vol. 1, pt. 2, chap. 4).
[10] Old Regime Parisian wages come from Baulant (1971) and Durand (1966). Most of the data for Paris were graciously given to me on tape by Philip Hoffman. Nineteenth-century Parisian wages are from Simiand (1932, Vol. 3, graph B, series f 16e).

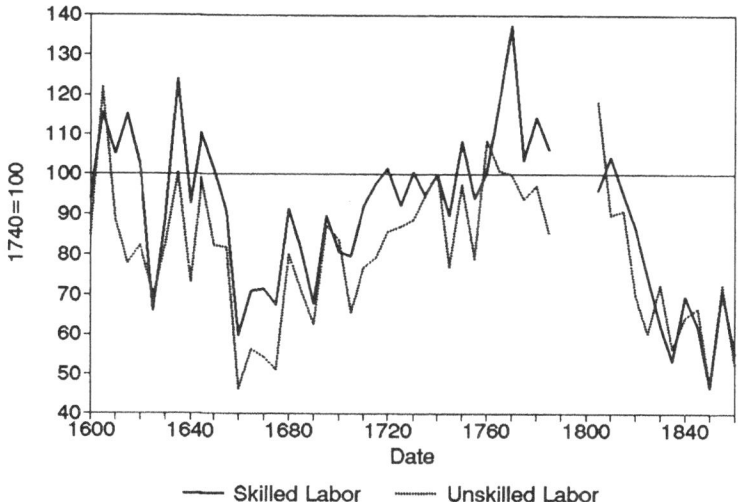

Figure 5.1. Wheat price–wage ratios in Provence, 1600–1860. *Sources:* Wheat prices from Baehrel (1962, Vol. 1, 647), until 1790. Nineteenth-century wheat prices from Labrousse, Romano, and Dreyfus (1970, 68–9). For sources on wages, see Appendix 1.

sented here were collected from a variety of archival sources by the author.[11]

Figures 5.1 and 5.2 present the price of wheat relative to wages for three areas – Paris, Provence, and Normandy.[12] While there are differences in these ratios in each area from year to year, one can discern a rough pattern that holds for most of France. Between 1600 and 1710, the ratios of wheat prices to wages fell irregularly. The trend was reversed after 1720, when a jagged upswing began. The price to wage ratios peaked at the time of the French Revolution. Relative prices then fell rapidly after 1789 and from 1820 to 1860 remained lower than those in

[11] Archival data were compared with published data for consistency. For Provence, see Baehrel (1962, Vol. 1, 604–13), Le Roy Ladurie (1966, Vol. 2, 1012–18), and Gangneux (1982, chap. 6). For Normandy, archival sources were completed with data from Desert (1977, 777). For further details on the construction of wage series for Normandy and Provence, see Appendix 1.

[12] The data come from a variety of sources. For wheat before 1789 in Paris, see Baulant (1968); in Provence, see Baehrel (1962, Vol. 1, 554). For consistency, these prices were compared with data from Le Roy Ladurie (1966, Vol. 2, 820–2). Levels are very close and the correlation coefficient is about .9. For Normandy, see El Kordi (1970, 282–7). All nineteenth-century wheat prices come from Labrousse, Romano, and Dreyfus (1970, *départements* of the Seine, Gard, and Calvados, 68–71, 190–1). To increase the representativeness of the series, wheat prices are centered five-year moving averages from the original series.

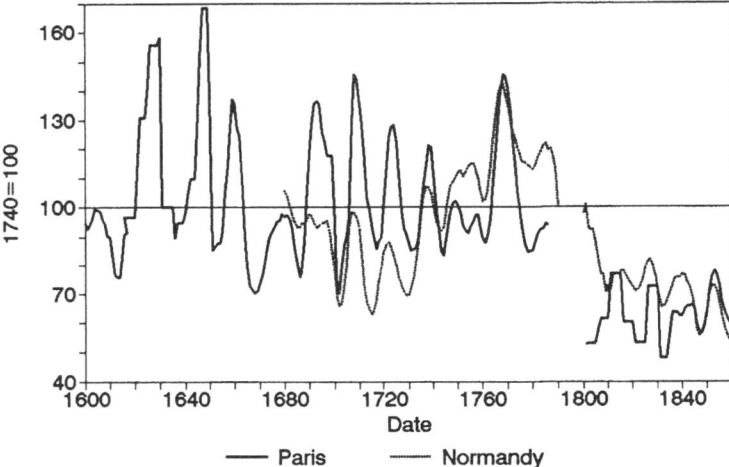

Figure 5.2. Wheat price–wage ratios in Paris and Normandy, 1600–1860. *Sources:* For wheat before 1789: in Paris, Baulant (1968); Normandy, El Kordi (1970, 282–7). For all nineteenth-century wheat prices: Labrousse, Romano, and Dreyfus (1970, 70–1, 190–1). For Paris wages: Baulant (1971) and Durand (1966). For nineteenth-century Parisian wages: Simiand (1932, 3: graph B, series f 16e). For sources on wages in Calvados, see Appendix 1.

the eighteenth century. While nineteenth-century expansion of drainage and irrigation thus occurred at an unfavorable time of high wages relative to commodity prices, the closing years of the Old Regime seemed particularly favorable to investment in land improvements despite the overall failure of the reclamation effort.

One crucial factor that such a crude estimate of relative prices ignores is technical change in agriculture. Indeed, if the productivity of land was substantially higher in the nineteenth century than in the eighteenth, the profitability of improvements may have risen while commodity prices fell. The ratio of wheat prices to wages is probably a good indicator of the price of land relative to the price of labor up to the middle of the eighteenth century because productivity increases were slow. After 1750, however, productivity increases and changes in output due to increased commercial activity diluted the relationship between land values and wheat prices. Thus, ratios of wheat prices to wages are probably unreliable indicators of the profitability of drainage or irrigation after 1750.[13]

Measures of the price of land that are more accurate than wheat prices would include land rents or land prices themselves. Land rents have been

[13] Traditionally, eighteenth-century agriculture has been characterized as technically stagnant, yet new evidence suggests that significant progress was being made. See Morineau (1970) and Hoffman (forthcoming).

Relative prices and the supply of water control

collected for the period between 1600 and 1789, while land prices have been collected from 1700 and 1860. Land rent data come from published sources and cover a broad geographic base. Unfortunately, however, these data do not provide any information for the period between 1700 and 1860.[14] Land price series, collected specifically for this study from archival material, cover the period from 1700 to 1860, but for only two small regions.

All previous studies of land rents have relied on data collected from lease contracts of religious estates. Most of these studies have focused on the question of whether real rent was rising or falling in the closing decades of the Old Regime. These data can also be used as a first approximation of changes in the relative prices of land and labor. Data taken from religious estates have the advantage of referring to the same tracts of land for two centuries. However, these data also have two disadvantages. First, no account is made of changes in the level of investment on land, such as buildings or other improvements.[15] Second, all the series stop in 1790.[16] Data for the nineteenth century are even harder to come by from published sources. As a result of these problems, it is possible to examine only how the relative prices of land and labor varied between 1600 and 1860 in two regions: Languedoc-Provence in the South and Normandy-Anjou in the West. Moreover, only price series for the South from 1600 to 1790 and for the West from 1670 to 1790 could be collected.

Five published land rental series were used to construct land price to wage ratios for southern France. Three series are from the Rhone River delta near Arles, one is from Montpellier, and one is from Lyon.[17] These five series run from 1600 to 1789 and cover both the period of successful drainage in the first half of the sixteenth century and the next 150 years, when little investment in water control seems to have occurred. To these published Old Regime data, one can add two new land price series for Cavaillon, a town about forty miles northeast of Arles. These series were collected from individual land sales contracts drawn

---

[14] There exists a series of rental data near Rouen for the period1735–1870 in Chaline (1966), but its behavior fits neither the eighteenth- nor the nineteenth-century patterns, so it seems preferable to exclude these data.

[15] Changes in buildings and other improvements were usually carefully noted in the estates leases, but most scholars dealing with religious estates simply ignored such issues.

[16] All the data come from the archives of religious estates that were confiscated during the Revolution.

[17] Two of the land price series come from Baehrel (1962, Vol. 1, 647). Another comes from Gangneux (1982, 144). This series traces the temporal revenues of the Order of Malta in Camargue, more than 90% of which came from land rentals. The series for Montpellier comes from Zolla (1893, 299–326, 439–61, 686–705). Finally, the data for Lyon come from Head-König (1972).

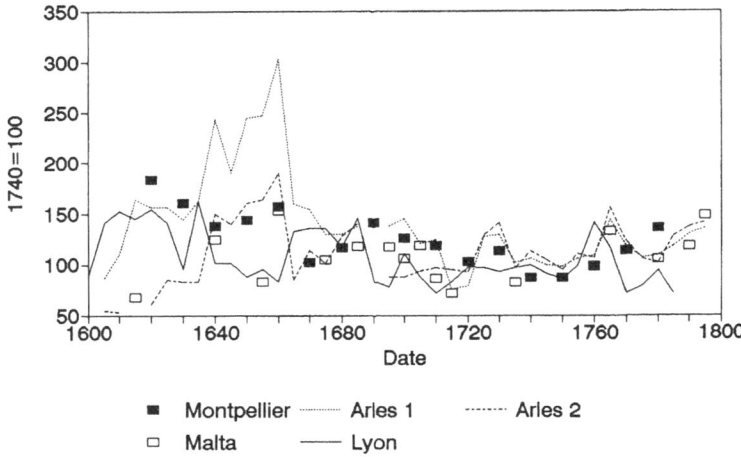

Figure 5.3. Land price–wage ratios in southeastern France, 1600–1789. *Sources:* Land price series for Arles 1 and Arles 2, from Baehrel (1962, Vol. 1, 647); for Malta, from Gangneux (1982, 144); for Montpellier, from Zolla (1893, 299–326, 439–61, 686–705); for Lyon, from Head-König (1972, 153–65). For wages, see Appendix 1.

up between 1700 and 1860.[18] All were deflated with unskilled wages from Avignon.

Figure 5.3 displays the normalized ratio of rental prices to wages for the seventeenth and eighteenth centuries. Generally, the first half of the seventeenth century was a period of rising prices of land relative to labor, while the middle and latter parts of the century were marked by a decline in relative prices. The decline in the relative prices of land and labor continued until the 1740s. In the second half of the eighteenth century, land prices increased faster than wages, suggesting that land was becoming relatively scarce. Nonetheless, there is substantial variation among the series, even those that are concentrated in the Rhone delta (Arles 1, Arles 2, and Malta).

The rental series suggest how sensitive land rental prices are to location even in the long run. Hence, investigating changes in the profitability of water control requires data for those areas where investment would have affected acreage. To this end, two land price series were collected for this study, one for dry land and one for irrigated land in the area where most of the investment in irrigation took place. Prices for both kinds of land were needed, because the profits of irrigation depended on the difference between the price for dry land and the price for irrigated land. As Figure

[18] For details, see Appendix 1.

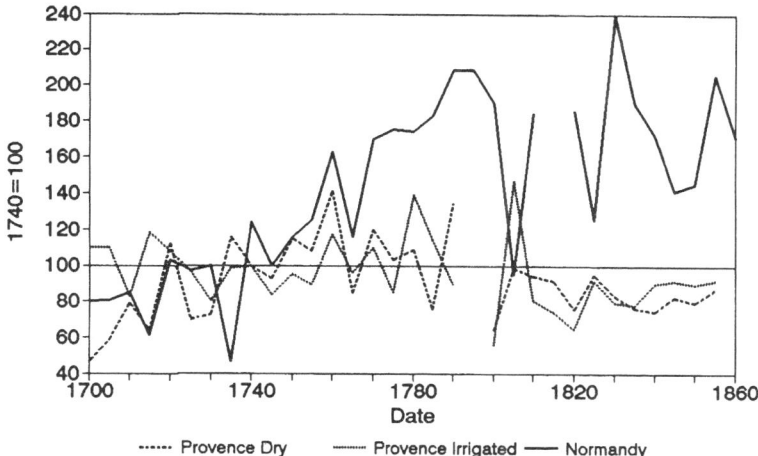

Figure 5.4. Land price–wage ratios in Provence and Normandy, 1700–1855. *Sources:* See Appendix 1.

5.4 shows, the price of irrigated land bottomed out in the 1750s and later rose steadily. The dry land data, however, suggest that relative prices were increasing over most of the eighteenth century. More important, prices of both dry and irrigated land did not rise sufficiently between 1780 and 1820 to keep up with wage inflation. Thus, during the nineteenth century the ratios of land prices to wages were less favorable for investment in land than they had been between 1740 and 1789. In short, southern relative prices were more favorable in the late eighteenth century than at any other time since the middle seventeenth century, and they do not suggest that changes in prices could have dramatically affected the returns to water control.

There are not as many rental series for the West of France as there are for the South. In addition, wage data are scant in this region for most of the period. There are, however, three published series of rental prices, two for Maine and one for Anjou, that range from 1670 to 1790.[19] Qualitative sources suggest that there was a substantial rise in rental prices from 1800 to 1820.[20] To these data, one can add a land price series that runs from 1702 to 1860, which was collected around the town of Troarn in Calvados. All these series are deflated by unskilled wage rates in Calvados. As with the southern series, those from the West, presented in

---

[19] One Maine series is from Zolla (1893, 423–64). The other Maine series comes from Garnier (1979). The series for Anjou comes from Michel (1978).

[20] Chabert (1945–9, Vol. 2, 117). Chabert suggests that, while rents rose by 40% from 1800 to 1820, there was enormous regional variation. In some areas, rents fell by as much as 10% while in others rents rose by 60%.

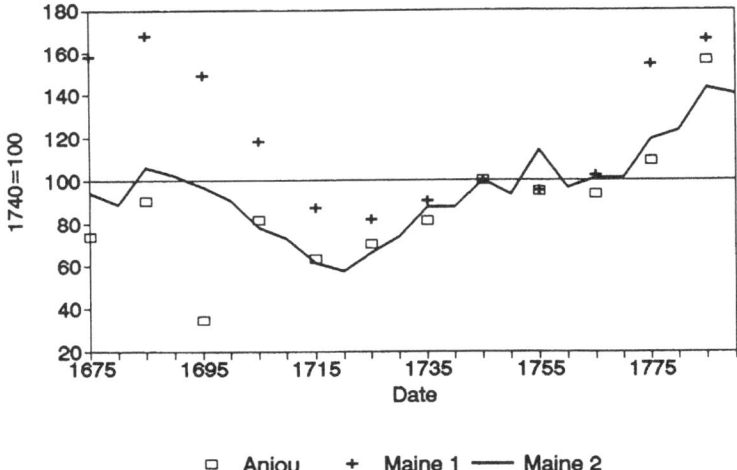

Figure 5.5. Land rent–wage ratios in northwestern France, 1700–1855. *Sources:* Land price series for Maine 1, from Zolla (1893, 423–64); for Maine 2, from Garnier (1979); for Anjou, from Michel (1978). For wages, see Appendix 1.

Figures 5.4 and 5.5, show a variety of patterns. The Anjou series displays signs of ever-increasing land scarcity over the period from 1670 to 1789, while data from the Maine region suggest that relative prices did not favor increasing investment in land until after 1720. The land price data from Troarn are even less optimistic, suggesting that relative prices may not have begun to rise until after 1740. All series indicate that the late eighteenth century was the most favorable period for investment in land since the late seventeenth century. Finally, the Troarn series suggests that land prices relative to labor prices were similar in the second half of the eighteenth century and in the first half of the nineteenth.

Over all of France, the data suggest that the profitability of improving land fell from the early seventeenth century to about 1730. After 1730, land may have become sufficiently scarce that the potential profits from improvement may have begun to rise again. Despite these broad patterns, however, heterogeneity dominates the relative price series. More important, while these trends in profitability correspond well with the history of attempted water control projects summarized in Chapter 4, they have little to do with actual successes. Indeed, if attempts at drainage and irrigation seem to be correlated with the relative scarcity of land, the success rate shows a dramatic discontinuity between 1789 and 1815. Before 1789, most drainage and irrigation projects were not even started, while after 1815 most were actually completed.

To explain the dramatic differences in the success rates of water control projects we can turn to three likely explanatory factors: interest rates, technology, and institutions. Because the relevant techniques and institutions differed between drainage and irrigation, their contributions will be detailed separately in the next two chapters. Here, we focus on the role of interest rates and the supply of credit in water control success. Conceivably, a significantly higher interest rate in the eighteenth than in the nineteenth century could explain why the development of water control had to wait until after 1800. In analyzing this possibility, the historian must decide whether to rely on the government's interest rate or that of private capital markets as the appropriate measure of rates in water control projects. One could rely on the government interest rate, which was above 6 percent during most of the eighteenth century and declined to about 4 percent after 1815.[21] Yet using the government rate might produce skewed results, since the substantial decline in government interest rates was not matched in private capital markets. In private markets, interest rates seem to have hovered around 5 percent during the entire period from 1700 to 1860.[22] Ultimately, it is this rate that is of relevance to water control projects because it was the rate at which large landowners – the principal backers of such projects – could borrow.[23] It is possible that institutional problems with credit markets constrained the supply of credit in the eighteenth century. However, both canal and drainage projectors were often backed by wealthy and powerful individuals who had access to abundant capital at a 5 percent interest rate.[24] Thus, interest rates probably cannot help explain the lack of investment in water control in the second half of the eighteenth century.

An examination of the historical evolution of land prices and wages suggests that late-eighteenth-century attempts to increase investment in water control and in agriculture in general did correspond to a period of increasing land scarcity. Hence, one should be wary of any attempt to classify the reforms of Old Regime administrators as pure redistribution. In fact, royal agents were responding to trends in the economy in their attempts to simplify the institutional mechanisms of investment in agriculture. To be sure, their focus on private property was, at least in part, ideologically motivated, and their preference for the nobility in part the result of a political calculus. Such ideological and political considerations

[21] See David Weir and François Velde, "The Financial Market and Government Debt Policy in France, 1750–1793" (Yale Univ., mimeo, 1990). See also Homer (1977, 223–4).

[22] See Jean-Laurent Rosenthal, "A Credit Market in Old-Regime France, L'Isle sur Sorgues, 1650–1788" (UCLA Dept. of Economics Working Paper, 1990) and Servais (1984).

[23] See, e.g., Postel-Vinay (1989).

[24] In any case, the possibility that the Revolution alleviated institutional constraints on credit would reinforce rather than weaken the argument presented herein.

probably influenced the choice of solutions to the problem of increasing agricultural output and so reduced the success of their reforms. Yet there is no way to deny that Old Regime France suffered a reduction in national output because of the failure to increase investment in land reclamation. Moreover, it appears that the loss in national product due to a frozen land supply rose from 1750 to 1789.

Investigating the role of either technology or institutions in blocking drainage and irrigation requires a change in focus – from the national to the local level – allowing a more detailed analysis. Legal reforms, the basis of institutional change, were rarely enacted nationally. Most often in the case of property rights, changes were negotiated region by region with judicial or fiscal officers who jealously guarded the *privileges* of each area. As noted earlier, each region's experience was the result of the interaction of its specific property rights with the centralizing tendencies of the state. In the cases of technology, different areas faced different problems. In addition, a given area's price were more responsive to the vagaries of the local economy than to national performance. The next two chapters address these issues, focusing in turn on drainage in Normandy and irrigation in Provence while leaving for later the tricky problem of deciding how these experiences illuminate the impact of institutions and technology on water control over the whole of France.

6

Drainage in the Pays d'Auge, 1700–1848: the weight of uncertain property rights

The importance of adequate drainage for agricultural productivity is well known.[1] In Old Regime France, drainage could have significantly increased agricultural output on two areas. First, in many regions of France, fields already under the plow would have greatly benefited from increased water control.[2] Second, as we have seen in Chapter 4, drainage played a crucial role in bringing new land under production. Between 1700 and 1850, draining fields already under the plow presented a complex institutional problem because it required reallocating customary, common, and private property rights at once. In the case of marshes, the problem was more modest, since only customary and common rights had to be altered. This chapter focuses on the problems associated with draining marshes in a specific area: lower Normandy in northwestern France (see Map 1.1). The climate and geography of lower Normandy make it ideal for a study of drainage because it endures both steady rainfall and inadequate natural drainage over a large proportion of its terrain. More specifically, within lower Normandy the area most in need of drainage was the Dives Basin (Map 6.1). The basin had originally been almost all marsh, but in the Middle Ages abbeys had spearheaded settlement and drainage. yet in the eighteenth century, much land still remained poorly developed; these lands are the focus of our first detailed study of the interaction between institutions and drainage.

The Dives Basin lies on the coast of the English Channel in Normandy, only a few miles east of Caen in the *département* of the Calvados. It corresponds roughly to the present canton of Troarn and is a very flat plain with small hills. Hydrologically, the Dives Basin can be divided between the areas south of the town of Troarn and the areas north of it. North of the town, the river Dives runs in a flat basin where the surface of the plain is near the level of the highest tides. The Dives has a difficult

---

[1] See, e.g., Allen (1982), who suggests that enclosures increased productivity principally because they were associated with drainage.
[2] Meuvret (1977–88, Vol. 1, pt. 1, 110–14).

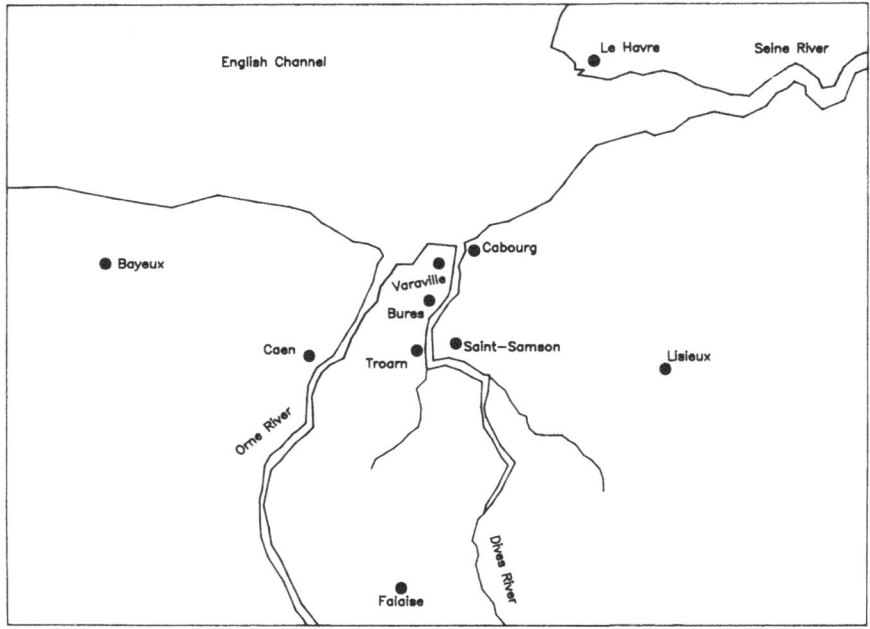

Map 6.1. The Pays d'Auge.

exit into the English Channel, and without human intervention it would periodically flood a large area north of Troarn. In addition, south of Troarn there are a number of marshes that also drain into the Dives.

All these areas would have witnessed some economic activity without any drainage. A significant portion of the land is composed of small hills, which did not require drainage. The wet marshes themselves were the locus of productive activities. Nonetheless, by the 1700s the owners of marshes had become convinced that reclamation through drainage would be a profitable operation.[3] They wanted to convert the marshes to year-round pastures and meadows, which they considered the most profitable way to exploit them. Year-round pasturing, however, required control of the entry and exit of water. During most of the year, a farmer would typically want to drain his grasslands, but in a particularly dry year or after mowing meadows, he might want to flood them. Such water control required the construction of levees and ditches.

In the nineteenth century, the marshes in the Dives belonged either to the state or to local communities. Individual peasants simply did not own marshes and rarely owned any pasture. Before 1789, theoretical owner-

---

[3] See AD Calvados, C. 4240–55. Between 1700 and 1789, it was proposed that almost every marsh in the area be drained.

ship was even more concentrated: Although some marshes lay in the hands of the king, communities, or various seigniors, most of the marshes – and a considerable amount of pastures as well – were owned by the area's largest seigniorial landlord, the Abbey of Troarn. Only near the sea, in the communities of Varaville and Cabourg, did the abbey fail to dominate the ownership of pastures and marshes.[4] Yet seigniorial ownership did not imply control, since a complex set of contractual arrangements between lords and villages reigned over marshes.

For the *conseil du roi* (the king's council), a committee that adjudicated property rights and issued drainage permits, the marshes in the area seemed, at least at first glance, to be devoid of economic value.[5] Thus, the complex fabric of use rights that regulated access to marshes was seen as a noneconomic remnant of feudalism that impeded drainage. The truth was in fact quite different: Marshes were valuable resources, and access to them was tightly regulated. Seigniorial owners derived small seigniorial rents from their marshes and also sold fishing rights, which were often economically significant. Moreover, because marshes were the water reservoirs of local mills, draining marshes could, by lowering the water level, have reduced the productivity of mills. Mills were the property of the seigniorial lord and an important source of revenue. Marshes were also of economic value to peasant communities because they provided mediocre but cheap pastures. The pasture they furnished was mediocre because of the lack of drainage, but it was cheap since the rent that the communities paid the seigniorial lords for pasture rights had often been frozen since the Middle Ages. In some cases, the right to pasture had become customary, and the communities paid the seignior nothing for their use.

Seigniors, government officials, and many farmers wanted to disentangle the web of contracts that ruled marshes. They wanted to transform the property rights system from a common to a private regime. In short, they wanted to divide and drain marshes. This would allow each farmer to invest in his piece of marsh and increase its value, and it would permit the lord to rent his portion of the marsh at a much higher rent than what he was receiving from villages.

Under the Old Regime, the overlapping claims of communities and seigniorial lords could be separated if both parties accepted the *triage* rule of 1699. The law made division and hence drainage extremely difficult. As noted in Chapter 4, in the case where lords retained clear property rights to marshes, *triage* (at least in Normandy) gave the lord one-

---

[4] Most pastures in the Dives area were marshes that had been drained in the Middle Ages. The provincial administration used ownership rolls to tax landowners for upkeep costs. See AD Calvados, C 4037–8.

[5] See AN H¹ 1495 for the opinions of the king's council and of the provincial administration.

third of the marshland and the community two-thirds.[6] In all other cases, dividing the marsh required the consent of the king, the lord, and all the villagers even though the lord may have actually controlled most of the marsh's resources. In any event, it is clear that, for villages and lords, settling questions of *triage* could raise obstacles to drainage. Conflicts over whether the *triage* rule could be applied to a marsh frequently provoked litigation and added significantly to the cost of drainage. These conflicts would, of course, disappear with the abolition of seigniorial property rights during the French Revolution.

In order to argue that it was the Revolution that removed obstacles to drainage, it is necessary to account for changes in drainage techniques between the eighteenth and the nineteenth centuries. Between 1700 and 1850, the techniques used for draining marshes, like those in the Dives Basin, were extremely primitive and did not improve. To be sure, Dutch and Italian engineers created complex drainage schemes with constantly improving techniques from 1500 onward, but in the case of small French marshes these sophisticated approaches were unnecessary. In 1850, as in 1700, the task was still a matter of digging ditches and putting up levees. Both of these tasks were accomplished by hand and demanded mostly unskilled labor. Floodgates did require skilled labor – carpenters and masons – and small amounts of building materials – wood and stone. Yet the major input was still labor, and there is no evidence of changing techniques in masonry or carpentry.[7]

Only after 1850 was the French industrial base sufficient to provide a new drainage technology based on concrete and steam power.[8] Before 1850, nearly all the work was accomplished by men, particularly unskilled men, with picks and shovels. Although there was certainly some learning by doing in drainage projects, the basic techniques remained the same. In the Dives Basin, nearly every supervising civil engineer from 1760 to 1850 formulated some proposal aimed at improving drainage. Over the course of the eighteenth century and the first half of the nineteenth, there was little change in what they proposed. From Remi Macquart in 1699 to Olivier, his counterpart of 1858, all the engineers who surveyed the area offered essentially the same solution to the water control problem. Olivier's proposal was an improvement on past proposals only because it considered the area as a whole and suggested both increasing drainage and monitoring how the increased flow of water in the

---

[6] See AD Calvados, C 4271.

[7] Although the financing methods varied from project to project, nearly all the expense of drainage was for laborers and masons. In the two cases where separate accounts were kept for floodgate construction and ditch digging, the cost of floodgates, including the labor to build them, was no more than a sixth of the total cost of the drainage network. AD Calvados, C 4262 and 6771, S 1269–1272.

[8] Desert (1977, 304).

river affected the rest of the area. Otherwise, nothing had changed. The peripheral drainage canals that Olivier proposed in 1858 were not new: They had first been proposed in the 1760s. Similarly, his plan to straighten the bed of the Dives dated back to the 1770s or earlier.[9]

The solutions to the problem of drainage that were proposed in the eighteenth century were thus little different than those proposed after 1800. It was therefore not inadequate technology that was responsible for the lack of drainage before the Revolution. To be sure, drainage projects were expensive investments, but again the problem was not their cost. Nor was it their revenues, for the evidence will indicate that drainage projects would have been profitable (had it not been for the cost of litigation) long before the Revolution. Rather, the obstacle was the lack of efficient institutions that would govern the distribution of the costs of a project among property rights holders.

―――――――

The argument that changes in institutions facilitated the drainage of marshes implicitly assumes that, in the absence of institutional barriers, drainage projects would have been carried out in the eighteenth century. In particular, more marshes would have been drained if the cost of resolving conflicts over property rights had been lowered. In other words, the validity of the hypothesis of institutional failure depends on whether or not the relative prices of land and labor would have made drainage of marshes a profitable operation in the absence of litigation. These issues were examined using data from two drainage projects for which sufficient evidence survives. In both cases, estimated rates of return strongly suggest that relative prices had little impact on profits.

To evaluate the profitability of drainage, let us return to the model presented in Chapter 5. Benefits will be calculated as the area drained times the price of arable land. While it would have been preferable to have a pastureland price series to evaluate the returns to drainage, pastures were too rarely traded in the eighteenth century to allow for reliable price estimates. Instead, our estimates rely on arable land prices. While some marshes were poorly drained and thus the value of the land may not have increased much, most often drained marshland was very valuable. Indeed, contemporaries agreed that drained marshland was the most productive kind of land, high-quality pasture. Thus, using arable land prices to evaluate the value of drainage gives a conservative estimate of the revenues associated with drainage – the sale of the drained land as pasture. To simplify the calculation, and because wages and land prices were relatively stable between 1700 and 1860 in the short run, all prices

―――――――

[9] Olivier and Sallembert (1856); see AD Calvados, C 6771, S 1004a, and 1004b for earlier projects.

will be valued at the time the project would have started (t). The revenue R_t from drainage in year t will depend on the price of land (p_t) and the surface area drained (n): $R_t = np_t$.

The issue of costs is more complex. A projector considering a drainage project in year t would confront three types of costs. First, the opportunity costs of the marsh (m_t); second, the cost of labor required to dig trenches and raise levees, where l_i is the number of man-days needed in each year while the network is being built. Third, since we are using an arable land price series, we must explicitly account for upkeep costs. Let upkeep cost be denoted by λw_t, where λ is the expected number of man-days needed to maintain the network each year. Both construction and upkeep costs will be assumed to be only labor costs. The estimated rate of return if the project is started at t and finished at T can be defined as Δ_t such that

$$\frac{np_t}{(1+\Delta_t)^T} - m_t - \sum_{i=0}^{T-1} \frac{l_i w_t}{(1+\Delta_t)^i} - \sum_{i=T}^{\infty} \frac{\lambda w_t}{(1+\Delta_t)^i} = 0.$$

We can test the profitability of drainage by estimating changes in Δ_t over time as a result of changes in relative prices.

To be sure, draining marshland did contain an element of risk, and we would expect the rate of return on risky investments to be higher than that on riskless assets. We must keep this in mind when we compare Δ_t with available interest rates for the period, such as those of the mortgage market. Except for the Revolutionary period, the mortgage rates remained stable at 5 percent, and although they were not without risk, they were the safest form of investment available to the eighteenth-century French. I also calculate benefit–cost ratios using both the mortgage rate and an estimated rate of interest constructed from more detailed French government *rente* data.

A rare eighteenth-century project's archives allows a deeper investigation into the cost of drainage. The records of Marais des Terriers, a marsh that was drained between 1714 and 1817, give detailed information about the returns to the individuals who headed the project. Computing rates of returns requires a series of assumptions. First, we assume that the only input was unskilled labor. Although there were skilled workers on all drainage projects, their wages are highly correlated with those of unskilled workers so that this assumption will not carry too great a risk of error. The assumption that unskilled labor was the only input allows us to convert the cost of the drainage network as well its upkeep into man-days by dividing the 1714 cost of drainage by the 1714 wage for laborers. The estimated wage bill in other years is then computed by multiplying this number of man-days by the wage rate in the year in question.

During the four years of the Marais des Terriers project, the projectors spent 44,000 livres.[10] I will assume that they spent equal amounts each year, or 11,000 livres annually. With wages for an unskilled laborer in the decade 1710–20 at 0.72 livre per day, the projectors used 15,200 man-days per year to build the drainage network. After 1718, the upkeep of the project ran at 2,000 livres a year, and since upkeep primarily involved maintaining the ditches and the main canal, it can also be regarded as a labor cost and converted into man-days. At the same rate of 0.72 livre per man-day, 2,000 livres is equivalent to 2,777 man-days.

Second, we make assumptions concerning the opportunity cost of the marsh that was drained. It belonged to the Abbey of Troarn and was drained under the technical direction of Remi Macquart, a royal engineer. After drainage, it was divided among the abbey, as seignior; the projectors of the drainage project, who were Macquart and four Parisian nobles; and the communities with rights to the marsh.[11] The abbey, which initiated the project, received one-sixth of the marsh, while the projectors received half as payment for draining it. The communities with rights to the marsh got a third of the drained surface as compensation for their customary rights. Neither the abbey nor the villages helped pay the cost of drainage. I will assume that the third of the drained marsh that the communities received was better pasture than what their limited access to the whole undrained marsh provided. In other words, I assume that the communities gained from drainage (after all, they did accept, without a legal challenge, the whole drainage project). There is good evidence that the communities were indeed better off after drainage. Though smaller, their pasture was greatly improved, for the undrained marshland had been flooded too often to offer pasture except for the summer months.[12]

Draining the marsh also involved a decline in the value of fishing rights

---

[10] AD Calvados, C 4073. Clearly, some of this money was used to resolve institutional problems. Assigning all of it to the physical task of drainage can only strengthen any findings of profitability.

[11] See AD Calvados, C 4295, for the original contract between all the parties of the drainage of the Marais des Terriers in 1699.

[12] According to AD Calvados, S 1270, the change in the value of private pastures from increased drainage was never less than 50%. The original drainage contract states that the land of the marsh had never produced anything but reeds and bad grass (AD Calvados, C 4295). Although the assumption that control of one-third of the drained land was better than limited access to the whole marsh, the estimation does not depend on it. Because I wanted to calculate the rate of return to the project for the projectors, I did not count in the revenue calculation some 300 arpents (150 hectares) that were the abbey's share (even though their drainage costs were counted); rather I used the cost of draining the whole marsh as my cost estimate. In effect, half the land drained was set aside to compensate for lost pasture, thus clearly biasing revenues downward.

and of a mill. Fishing must have greatly declined, although it is clear that commercial fishing continued in the drainage ditches. The mill suffered a 50 percent loss of power. Although neither milling nor fishing completely disappeared, I make the assumption that drainage brought both activities to an end. That assumption should bias the estimated rate of return downward. In fact, the rental value of the mill was 1,000 livres a year and remained so throughout the period 1650–1766, despite the completion of the drainage project in 1717. Hence, although the rental price did not increase in a period of general inflation, it did not decline, a clear indication that damage was limited. The fishing rights seem to have generated about the same revenue, but there is no certainty that the archives of the Abbey of Troarn contain the full set of rental contracts for any year. The best estimate is that those rights represented another 1,000 livres of yearly revenue. At worst, then, the loss of fishing and milling rights would have amounted to 2,000 livres per year. Typically, assets like milling and fishing rights could have been purchased for a price very close to the capitalized value of the rent they earned. This can be estimated using the rate of interest – either the 5 percent mortgage rate or the rate estimated from British data. At the 5 percent rate, 40,000 livres is the estimated opportunity cost for the marsh from the capitalized value of the rents earned by the abbey from milling and fishing.[13]

Although Macquart's widow and three of his associates sold their shares in the Marais des Terriers to the remaining projector, Oursin, the sales contracts were impossible to trace. Thus, we do not know how much the drained marsh was worth, but we do know the surface area of the projectors' share. Revenue estimates are computed assuming that the projectors' 450 hectares were sold at the prevailing average price for naturally drained land minus the present discounted value of future maintenance costs. Again, the use of such an average price for land understates the revenues, since drained marshland was reputed to be the best pastureland in Normandy.

To estimate the discounted value of the future maintenance costs, I will assume that, having drained the marsh in four years, the projectors sold the land and created a sinking fund to deal with maintenance costs. This sinking fund, I will assume, was composed of bonds (*rentes*) with a yield equal to the interest rate. With these assumptions, the capital cost of future upkeep will equal the wage costs for 2,777 man-days of labor divided by the interest rate. (In the case of the fixed interest rate it will

---

[13] Most of the data on mills comes from the AD Calvados, H 8160–6 (Abbaye de Troarn). The archives of the abbey contain rental contracts for the mill from 1665–1760. The rental price remained nearly 1,000 livres per year even in the contracts for 1710 to 1730, when the abbot sued the projectors arguing that his mill had lost some of its power.

Table 6.1. *Average hypothetical profits for Norman drainage projects*

Period	Annual rate of return (%)		Benefit–cost ratios				Interest rate (%)
			Mortgage rate		Government rate		
	Terriers	Troarn	Terriers	Troarn	Terriers	Troarn	
1702–50	37	47	2.0	1.1	2.2	1.2	5.4
1754–86	69	225	4.3	2.3	3.9	2.1	6.5
1790–1814	85	258	5.5	2.6	4.8	2.2	7.8
1818–50	78	184	4.2	1.9	4.2	2.0	5.0
1854–70	74	127	3.6	1.5	4.0	1.7	4.2

Sources: See Table A1.2.

be 55,555 times the wage.) The expected rate of return Δ_t on the project if it had been started in year t thus solves

$$\left((450p_t - \frac{1}{r_t}2{,}777w_t)\frac{1}{(1+\Delta)_t^4}\right) - \left(15{,}200w_t\sum_{i=0}^{3}\frac{1}{(1+\Delta_t)^i}\right)$$
$$- 2{,}000\frac{1}{r_t} = 0,$$

where p_t, w_t, and r_t are the price of land, the going wage, and the interest rate in year t, respectively.

The second, fourth, and sixth columns of Table 6.1 give the hypothetical rates of return for the Marais des Terriers broken down by periods.[14] Despite the conservative assumptions, all of which should bias the estimates downward, the project returned at least 20 percent per annum, assuming that the project was completed in four years.[15] As shown in Figure 6.1, except for 1734 – a year for which the estimate of the price of land is unreliable – between 1718 and 1789, the project's internal rate of return was never less than four times the rate of return on mortgages or government securities. Benefit–cost ratios, estimated using either the 5 percent mortgage rate or the government interest rate, tell the same story; except for 1734, these ratios are above 1.5 (see Figure 6.2). Thus, regardless of what interest rates are used, significant profits could have been earned from this project, all the more so since the conservative assumptions probably lead to an underestimate of the rates of return. Moreover, although all the estimates for the eighteenth century are lower

[14] Table A2.1 gives the complete series of estimated rates of return.
[15] The project would have remained profitable overall if it had lasted five years instead of four. Of course, profits would be lower, but one could similarly alter profits by revising the value of drained land or the cost of maintenance or even the amount of labor required in each year.

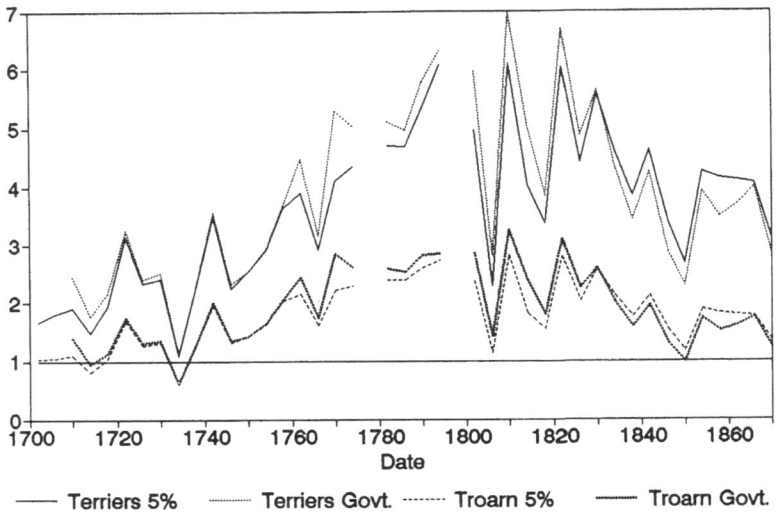

Figure 6.1. Hypothetical benefit–cost ratios for Norman drainage projects. *Sources:* See Appendix 2.

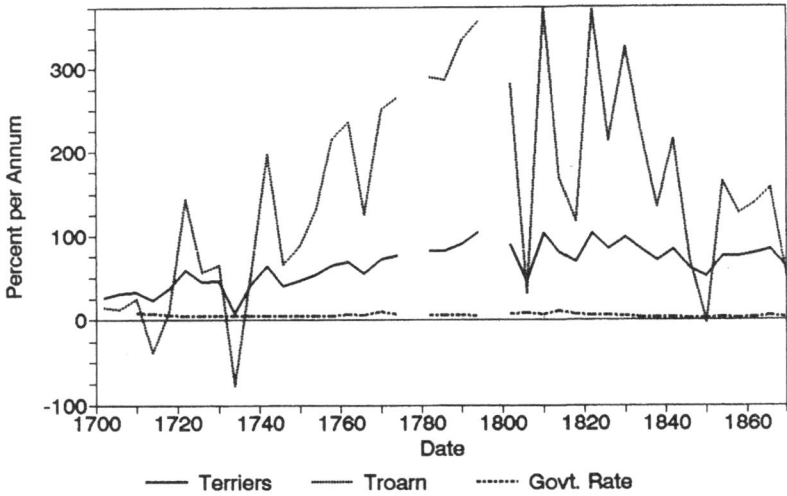

Figure 6.2. Hypothetical rates of return for Norman drainage projects. *Sources:* See Appendix 2.

than those for the nineteenth century, they leave us with the same conclusion: If the Marais des Terriers is any evidence, drainage ought to have been profitable as early as the 1720s.

A more convincing test of the market failure hypothesis involves computing the hypothetical rates of return to drainage projects that occurred in the nineteenth century. In the 1820s, the government provided institutional mechanisms for the division of common lands that ultimately facilitated the drainage of marshes. The measures affected marshes because the Revolution transformed the marshes, which had been seigniorial property, into common lands owned by villages. Like other common land, marshes were divided equally between each household in the village. The entire operation was supervised by the government, and before the marshes were divided, a basic drainage system was put in. Households in the village had to pay for the share they received, and the price they paid was set to compensate the community for revenues it had collected from the marsh and to pay for the drainage network. Sufficient data are available from one such division of a common marsh – in Troarn in the 1840s – to estimate a crude rate of return.[16]

In the division of the common marsh in Troarn, villagers received an average parcel of 0.3 hectare for the price of 100 francs. I will assume that the entire 100 francs went to pay for the drainage network. This assumption no doubt exaggerates the cost of the network because some of the money paid by the villagers reimbursed the village for lost revenue from the marsh and some went directly into the central government's coffers.[17] In Troarn, these revenues came from pasture rights (there was no mill on the marsh), but we will use other evidence to estimate their magnitude. I shall also assume that yearly upkeep costs amounted to 5 percent of the cost of the drainage network. Actual upkeep costs for Troarn were not available in the archives, but account books of other drainage projects suggest that this is an appropriate figure.[18] The cost of both the drainage network and its upkeep are estimated as labor costs.

The most difficult cost to estimate is the opportunity cost of the pasture forgone by the villagers. Since undrained marshes were never sold, one cannot use sale prices. However, contemporary sources (the tax service, the state, and dividing communities) did estimate the value of marshes from their revenues before division. These estimates average 1,100 francs per hectare, or 46 percent of the average price of land between 1838 and 1850. We will assume that the marsh in Troarn was worth slightly more than this – 50 percent of the average price of land. Again, this is a conservative assumption, for the Troarn marsh had no mill, and some of the value of the pasture forgone was actually part of the 100 francs per par-

---

[16] In the seventeenth century, the seigniorial owners of the marsh at Troarn had attempted to exercise *triage* and drain the marsh. In the eighteenth century, the communities that then owned the marsh also attempted to have the marsh drained. Yet as we shall see later, litigation prevented the realization of any drainage. AN H[1] 1692.

[17] Ponteil (1965, 31).   [18] AD Calvados, S 1269 (syndicat de la Dives).

cel that we assumed paid for the drainage network. Once again, we have biased the rate of return downward.[19]

The project was carried out in only one year, a fact that makes the calculation of rates of return very easy. As the third, fifth, and seventh columns of Table 6.1 indicate, this project also would have earned significant profits independent of the time it was carried out. Like the Terriers project, the Troarn project would have been very profitable in the eighteenth century. The average rate of return for the eighteenth century (1702–89) was above 100 percent. For the period 1750–86, when drainage became an important policy issue, the rate of return is greater than twenty times the interest rate on government bonds. The very high rates of return come from the fact that it was a small project that could be carried out quickly. Although in an absolute sense the Terriers project was more profitable than that of Troarn, its returns were delayed for four years, reducing the rates of return. Clearly, some marshes that were not drained before the nineteenth century could have been profitably drained fifty or one hundred years before.

Both estimated rates of return and benefit–cost ratios (displayed in Figures 6.1 and 6.2) were calculated in a conservative fashion, yet both lead to the same conclusion: Without litigation costs, drainage projects would have given a projector sizable profits. With the exception of one year, the estimated rates of return for both projects in the eighteenth century exceeded 5 percent – the mortgage rate – by a wide margin. Of course, one could imagine that even higher rates of return would have been necessary to compensate eighteenth-century projectors for the risks they took. Yet risks were actually limited, for the marsh returned its value in case of failure. Furthermore, the technology involved with drainage was simple. Typically, a drainage network failed, as did the one in the Marais des Terriers, not because the network failed to drain the marsh, but because of lack of upkeep.[20] In any event, the estimated rates of return for drainage projects in the eighteenth century seem high even for risky ventures. They were almost always well above 30 percent, a figure that compares very favorably with the estimated 18 percent rate of return for contemporary French slave trade ventures, which were notoriously risky.[21] Finally, because these rates of return were constructed in a con-

---

[19] One problem with using such drainage projects as measures of profitability is that commercial exploitation started only after drainage took place. The exact specification of the estimation for rates of return is

$$\left(\left(0.1505\,p_t - \frac{2.73}{r_t}\,w_t\right)\frac{1}{1+\Delta_t}\right) - 54.7\,w_t = 0.$$

[20] AD Calvados, H 8163.
[21] See Forster (1980, 5 and note 7). Stein (1979, 37–45) suggests that average profits in the slave trade were in fact much lower, at least in the 1780s.

servative fashion – to produce high costs and low returns – it seems safe to assume that the actual rate of return would in all likelihood have been even higher in the eighteenth century. The conclusion that the market for drainage was very inefficient in the eighteenth century thus seems inescapable. The fact that a great deal of attention was focused on drainage in the later eighteenth century suggests that the problem was not due to a lack of economic acumen among Norman landowners. Rather, marshes remained undrained because of a market failure.

———

Eighteenth-century administrators and investors were well aware of the potential for high returns to drainage and the relative lack of drainage. The most important administrators in drainage affairs were the *intendants,* the king's special representatives in a particular area called a *généralité*. In Normandy, the *intendants* supported drainage schemes even though they were well aware of the difficulty of carrying such projects through. In 1770, the *intendant* of Caen, Esmangard, remarked in connection with marshes not far from the Dives:

The marshes have, in their present state [undrained], produced nearly no revenue, even though the soil is good and could become through drainage one of the most fertile of the region. . . . There is no kind of obstacle a portion of the village has not created to prevent the division of the marshes and their drainage. . . . The Count of Langeron and the Marquee of Lambert [the promoters of a project to drain the marshes in question] have been progressing only step by step because everything has been done under duress and the different legal procedures that have become indispensable have brought about considerable delays.[22]

Esmangard's attitude exemplifies the interest that *intendants* had in drainage and also their inability to accelerate judicial procedures. Developers shared their belief in the potentially high return to and high social value of drainage. A would-be projector concluded in 1765:

An operation conducted with such generosity should necessarily return precious land to agriculture. The important affair is to drain some land, an operation deeply desired in a province such as lower Normandy, where the inhabitants suffer much from stagnant water.[23]

---

22 An H¹ 1496, fl. 62, September 17, 1770. A similar example comes from the South, where the *intendant* of Montpellier discussed the drainage of a marsh in 1760. After approving the project from an economic and a technical point of view, the *intendant* came to the issue of property rights. He discussed the demands made by the projectors for eminent domain privileges but concluded that the privileges requested by the projectors were probably too weak to ensure realization of the project (An F 10 318).

23 AD Calvados, C 4200, fl. 62, 1780. That local communities or landowners were often opposed to drainage has surprised a number of historians. Staunch opposition, however, could result in higher levels of compensation. In many cases, the projectors attempted to avoid compensating communities for customary rights. See, e.g., Forster (1980, 77–8) and AD Calvados, C 4203, June 1765.

He went on to discuss the problems that he had faced in finding land that could be drained without litigation.

Intendants and projectors worried about litigation because the distribution of property rights under the Old Regime exacerbated the problems associated with drainage. The difficulties that drainage would pose in any market, even those in which property rights were certain and litigation was not an issue, are obvious. Water control enjoys very significant positive externalities and economies of scale. For example, it would have been impossible to drain part of the Marais des Terriers without draining it all. Thus, it was necessary to get all the owners of the marsh to help pay the cost of drainage. Furthermore, drainage imposes negative externalities on all lower lands, for each new drainage project forces owners of lower lands to raise their levees. If the drainage of a whole hydrographic basin is contemplated, this issue is largely irrelevant, because the projectors will have to take into account the costs associated with the negative externalities. In France, however, the scale of suggested projects was much smaller than whole hydrographic basins. In such a situation, the institutions that allocate the costs of externalities among landowners are crucial for the successful development of any drainage project. In Old Regime France, conflicts between projectors and owners of land affected by drainage were handled judicially, and there was no clear precedent to lead the courts to rule that projectors were liable, say, for downstream damages due to increased water flow. Thus, downstream landowners were often opposed to projects from the start, and their opposition significantly raised the cost of drainage.

When there are many participants in a project, the costs of drainage also depend on the institutions that resolve conflicts among participants over the distribution of costs and benefits. If marsh ownership is sufficiently dispersed, no drainage can occur without a rule to allocate costs among owners. Moreover, whether the development is undertaken by someone exterior to the group of landowners, by some subset of the landowners, or by the landowners collectively, a rule of profit division is necessary. A projector who undertakes the project will have to be paid, and the landowners will have to divide the remaining benefits. Clearly, the rule that transfers a portion of the benefits from the landowners to the projector will determine the profitability of the projects for the projector. Division rules thus become important parameters in the equilibrium quantity of land drained, even though they do not affect the social value of the project because they address purely redistributive issues.

In the case of well-defined property rights and certain rent valuations, a simple institutional structure (having the state set the rule for sharing costs and benefits, for instance) could presumably have avoided much litigation. In the case of eighteenth-century property rights over marshes, however, litigation was almost inevitable because of informational asymmetries. The price of the property right in question – pasture rights over a marsh slated for drainage, for example – was in fact private information to the owner because he, better than anyone else, knew the nature of the marsh, the quality of its pasture, the problems of flooding, and thus the profits it could be expected to generate. None of this information was truly private, but clearly it was expensive to acquire, especially in the case of marshes that were generally not sold. Landowners obtained this information readily because they observed the returns of using or renting out their property rights to the marsh, but the state or a projector trying to reimburse a landowner for the loss of the property right would have had difficulty estimating what the undrained marsh was truly worth. Even if the projector used unbiased methods to estimate the costs and benefits of improvement, it would be difficult to devise a compensation rule that would leave all property rights owners no worse off than they would be with the marsh undrained.[24] When property rights owners were improperly compensated, they could appeal the contract in court. Hence, costly litigation was also likely because projectors and the owners of property rights to a marsh had access to different sources of information about the value of property rights.

Information problems also affected the compensation of negative externalities associated with drainage. For example, the value of damages to a mill that suffered from drainage was better known to the owner of the mill than to the projector. The mill's rental contracts were private information; hence, any damage suffered was also private information. Thus, the uncertainty in the value of damages or benefits was yet another cause for litigation, adding to the cost of supplying drainage.

Another and probably more important source of litigation was overlapping or uncertain property rights. Here the projector faced yet another problem: whom to compensate for damages. Litigation was often the only way to determine who owned the right to compensation. Under the Old Regime, a large number of individuals often had some title to the same piece of land. The list could include landowners, farmers, religious institutions, seigniorial lords, or the state. In Normandy, where marshes were most often common lands, claims to the property were often divided among the lord, the king, and the community. Because drainage was quite costly, projects could not start until all uncertain property rights had been clarified, and the attribution of property rights over marshes was the source of numerous suits.

The supply of drainage thus depended not only on such normal market forces as the relative prices of land and labor or the interest rate, but also on institutions. Because it was a public good, drainage required rules for

---

[24] Wiggins and Libecap (1985) offer another example of the costs of imperfect information when there are increasing returns to scale.

the division of costs and benefits. Individuals had incentives to litigate to reduce the losses they faced from the project or to increase the benefits they might receive. Uncertain property rights made it even more difficult to avoid litigation. Only under an institutional regime that resolved these difficulties was drainage likely to proceed, but in Old Regime France institutions failed to do so. As we shall see, it was not until after the Revolution, when institutions were completely changed, that drainage could be successfully pursued.

The institutional structure that governed drainage in the eighteenth century was primarily judicial. Conflicts were resolved in court. Two types of institutions governed litigation over property rights. The first were regional rules about access to common or seigniorial lands. The second were royal laws. Conflicts over either regional rules or royal laws, as well as over royal permits to drain, were handled in royal courts, and because all marshes were in fact common lands, the rules of *triage,* which I described earlier, generally prevailed. The only exception concerned royal land. While, technically, royal marshes could not be sold, rights to them could be leased in perpetuity. Royal land was particularly attractive to projectors, because the property rights questions concerned only the king.

Yet projects trying to reclaim land faced litigation independent of who the feudal owners were. The primary cause of litigation over marshes in the eighteenth century was conflict over property rights. By 1700, communities and seigniorial lords had accumulated overlapping property rights to most marshes. Drainage, however, required that marshes be divided and that the overlapping property rights be sorted out for the allocation of costs and benefits. In practice, marshes were at the center of three sorts of litigation. The first pitted king against seignior over the determination of seigniorial property. At issue was who held the seigniorial rights to the marsh. In Normandy, the seignior almost always won these suits, because, as we saw earlier, few marshes were part of the royal domain. The second type of litigation involved seigniors and communities and focused on whether the *triage* rule was applicable to specific marshes.[25] The third type of litigation – within the communities – was due to conflicts over the division of the surplus when the communities had strong use rights to the marsh.

In areas where villagers had strong use rights to marshes, access was

---

[25] In the case of the Marais des Terriers, the abbey was seigniorial lord and owner and thus had one-third of the marsh. In the case of the marsh of Troarn, it was seignior of Troarn but not the owner of the marsh. The abbey's share of the marsh would thus depend on what division rule was chosen. AN H¹ 1496 and AD Calvados, C 4293. For a contemporary view, see AN H¹ 1496, Marais de Chaumont.

determined by local custom; changes in access to marshland required changing customary law. Individual villagers had standing before the courts if they wished to appeal drainage projects because marshes were ruled by customary law. Drainage changed access to marshes by dividing the improved land into parcels for private rather than communal use, so villagers could credibly argue that their customary rights were violated. Since customary law as part of the *privileges,* royal courts were unable to dictate reform of customary law. As a result, if communities had strong use rights, the unanimous consent of all villagers involved was necessary, for villagers could always sue. Faced with appeals by landowners, royal courts neither enforced the drainage grants nor decided on a new allocation of the drained land. As a result, there was no judicial finality in conflicts over drainage grants.

The suits almost inevitably ended up in the king's council, the highest royal court. Drainage projects involved the high court even in the absence of suits because they changed the flow of water, and thus required a royal grant from the council. Royal grants were, at least in theory, contracts that divided costs and benefits among the different parties in drainage projects. They also awarded the monopoly right to carry out projects to a specific individual called the projector. Except for legal costs, these grants were supposed to be free; however, projectors were well aware of the need for political influence in order to obtain a speedy and favorable verdict from the council. In the case of the Marais des Terriers, for example, Macquart, who was the originator of the drainage project, found four Parisian nobles who became his associates and who presumably had some influence in the king's council.[26] The abbot of Troarn, the seigniorial lord of the Marais des Terriers and the opponent of Macquart in nearly every suit, wielded considerable influence himself.

Anyone dissatisfied with a proposed royal grant could oppose it by filing a brief with a royal court. Although technically an appeal of a royal decision, the brief took the form of a suit against the projectors. The suit could be pursued before the *baillage* (the local court) or the Parlement (the regional royal court of appeals) in Rouen and then go back to the king's council for another appeal. If the opponent found the prospects of the regular court system unsatisfactory, he could file with the Eaux et Forêts courts, the special water and forestry judicial system. Indeed, in the case of marshes the jurisdiction of the regular court system overlapped with the jurisdiction of the Eaux et Forêts. Here, too, final appeals could always be sent to the council. Any court that accepted an appeal would automatically grant a staying order, thus preventing the realization of any drainage network until all suits had been resolved. The pro-

---

[26] AC Vimont (Saint-Pierre Oursin), 9 E 761/75.

cess of court enforcement of drainage grants was very complex, lengthy, and expensive, and it could delay a drainage project for a long time, often several decades.

Old Regime royal governments were well aware of the severe economic costs associated with court delays. A number of ministers attempted to centralize judicial authority not only to reinforce royal power, but also to reduce delays. Beginning in the 1740s, some new drainage grants issued by the king's council contained a clause that stated that they could be reviewed only by the *intendant* or the council itself. Unfortunately, because such direct appeal clauses were not automatically included in all new drainage grants, the effects of this reform were limited. Projectors had to request the privilege of such a direct appeal clause in their grant. Soon, however, almost all projectors began to seek direct appeal clauses. As de Blossac, the *intendant* in Poitier, remarked:

Such an edict [that only allowed appeals directly to the council for a particular drainage project] would not be unique in its kind, and all those who have started drainage projects have obtained similar ones so they could avoid the length and formalities of the procedures that are nefarious to enterprises of this nature.[27]

The result was that the king's council became overloaded with appeals, and they took longer and longer to process. In Normandy in particular, this reform had little impact. Unlike other provinces where the *parlements* and other local institutions played important economic and judicial roles, Normandy was in the domain of direct royal power, at least as far as drainage was concerned. Indeed, as early as 1711, many Norman court cases over marshes quickly reached the king's council.[28] The centralization of judicial authority in the council did not facilitate drainage in Normandy. After 1740, litigation before the king's council took at least four years to reach a verdict, which was never enforced. Indeed, the council was always willing or compelled to reexamine past decisions. Judicial delays affected all the drainage projects of which we have records, for all were in some way involved in litigation before the council. It is no wonder, then, that most of the numerous drainage projects proposed in the eighteenth century were failures.

Three examples illustrate the problems with litigation that thwarted many eighteenth-century drainage projects. The first example is the marsh of Troarn and all the unsuccessful attempts to divide and drain it. The legal battle over property rights to this marsh started in the seventeenth century and was still raging a hundred years later, when the Revolution

[27] An H¹ 1497 (1776).   [28] AD Calvados, C 4293.

finally brought the litigation to an end. In a first phase, the abbey attempted to drain the marsh in order to convert it to pastureland and arable land.[29] The four communities (Troarn, Saint-Samson, Saint-Ouen-de-Bures, and Barneville) that enjoyed use rights over the marsh sued the abbey, arguing that their rights were customary and that the abbey had no right to drain the marsh. The communities won their suit in the king's council in 1667. The abbey's claims of ownership were invalidated, and the abbey was forced to destroy whatever drainage it had begun. The four communities now enjoyed strong property rights over the marsh. The verdict implied that any future division would require the consent of all villagers. What was not resolved by this trial was the division of property among the communities and the abbey in case of drainage.

A second phase of litigation erupted in 1711 when the same communities attempted to divide the land among themselves and to drain the marsh. A proposal was put forward to share the marsh on the basis of population. Although the king's council was solicited twice for a division edict, no cost-sharing rule could be agreed upon. The *intendant* was either unwilling or unable to enforce the decision of the council. The project was once again abandoned. In the 1770s, the issue of drainage surfaced again, and the battle over which method should be used to divide the marsh was renewed. The three smaller communities demanded equal shares, whereas the larger community of Troarn stuck by the 1711 proposal. Although Troarn later abandoned its position, the case was still in the courts at time of the Revolution. Clearly, what kept this marsh from being drained was not an issue of profits or of technology, but rather the absence of any precedent or other rule to divide the land among the different title holders. In fact, after the Revolution, when the institutional structure had changed, the marsh was divided among the communities, drained, and then split up among the villagers.

The second example involves the Marais des Terriers.[30] Although it was ultimately drained, the project provides a telling illustration of the length and extent of litigation. The original contract between the projector Macquart and the Abbey of Troarn dates from 1699, but the royal grant permitting drainage was not enacted until 1711. By 1711, Macquart had formed a partnership with four Parisian nobles, who provided capital and political influence while he offered technical ex-

---

[29] Drainage attempts are documented in the archives of the central government (AN H¹ 1492), the *intendance* of Caen (AD Calvados, C 4263, 4293), and the communities (AC Bures, Marais).

[30] AD Calvados, H 8163–6, C 4293, and AC Vimont, 9 E 761/75. There were actually a number of suits over the question of maintenance. It is unclear whether the network of ditches and canals was improperly maintained or whether the abbot attempted to renege on his financial obligations.

pertise. Despite a legal contract reinforced by a royal grant, Macquart and his associates became involved in suits with the abbot of Troarn. By 1717, the king's council was examining two different suits brought by the abbot. In the first, the abbot attacked Macquart and his associates, claiming that his mill was less productive as a result of the drainage they had undertaken. The abbot argued that both the royal grant and the original contract between himself and Macquart made provision for leaving the waterfall of the mill intact. The second suit, also between the abbot and Macquart, concerned the quality of drainage. The abbot won the first suit and was awarded 1,000 livres as damages. He was also given the opportunity to sell his mill to Macquart and his associates at the capitalized value of the rent before drainage affected the water flow. He lost the second suit over the quality of drainage, and work on the drainage network continued. In 1723, the network was declared complete, and as stated in the grant of 1711, each landowner was to be taxed for upkeep costs according to the surface he owned. To ensure compliance with the maintenance provision of the 1711 grant, a new contract that provided for the distribution of future maintenance costs was written between all the owners of drained land and Macquart and his associates.

The legality of this new contract was tested in 1726 when the abbot was back in court arguing that the projectors (Macquart and his associates) had broken both the 1711 and the 1723 contracts because they had inadequately maintained the drainage network. The *intendant* agreed with the abbot that work had to be done, but he also forced the abbot to pay his share of the costs, which the abbot had hitherto resisted. Despite the customary appeal by the abbot, this was one of the rare occasions when the king's council refused to review the *intendant*'s decision. Litigation, however, was not over. In 1740, one of Macquart's associates, who had bought all the other projectors' land, a Monsieur Oursin, brought suit against the abbot. Oursin appealed to the council because the abbot had raised the floor of the drainage canal to increase the fall at his mill. Again, the abbot lost, but the projectors of the marsh faced other suits from seigniors who claimed rights to the marsh and who attempted to impose seigniorial dues or exercise *triage*. Although Oursin managed to win all of these trials, the legal costs were quite large, and there is only one mention of reparative damages.

The third example comes from the village of Ranville, a parish that spent 645 livres between 1747 and 1748 to defend its lease on a marsh.[31] The marsh was part of the royal domain (the property of the king), and domain officials attempted to break the village's lease so as to raise the

[31] AD Calvados, C 4288, and RC Ranville, Marais.

rent on the land. The villagers won their suit only to face a renewed threat in 1856 when domain officials leased the marsh to a projector named d'Avenel. The village opposed the grant on the basis that their tenure had become customary, but it lost on appeal to the king's council. The council wanted to encourage drainage and leased the marsh to d'Avenel. He promised to drain the marsh, while the village wanted to use it as common pasture. D'Avenel, however, never even began to drain the marsh. In a final suit waged during the Revolution, the villagers argued that only the corruption of an *intendant*'s delegate had allowed d'Avenel to take over the marsh justly and that it should be returned to them. This case dramatically demonstrates that not even royal property rights were sufficiently well established to avoid litigation.

The complication, length, and dates of the lawsuits make it clear that the Old Regime had failed utterly to limit lawsuits over the property and administration of marshes. The problem worsened in the 1760s when fiscal difficulties forced the monarchy to increase the revenue from the king's domain. As a result, leases on royal possessions, including marshes, were revised upward. This revision led to renewed attempts to auction leases to royal possessions instead of renewing the old contracts. D'Avenel was able to rent the marsh at Ranville in such an auction. Royal officials were particularly fond of schemes like d'Avenel's, because they would raise the value of the royal domain without any royal investment. In the long run, drainage would increase the value of the marsh and increase the rent the state could earn. As the case of the marsh at Ranville suggests, royal interference often complicated property rights and increased the level of litigation over marshland, without promoting drainage.

During the late eighteenth century, perhaps because of increased fiscal needs, the central government became more concerned with promoting economic growth. In order to increase the cultivated acreage, the government enacted a series of reforms to encourage drainage. These included reforms of the judicial system, subsidies, and a streamlining of the process of granting marshes to projectors. All of these measures should have lowered the cost of drainage, but the economic record of the reforms was mixed, at best.

Consider, for example, the judicial reforms. As we have seen, the attempts at judicial centralization did not reduce – and perhaps increased – the delays associated with litigation over drainage. The delays, of course, had a significant impact on the profitability of the ventures. Moreover, the amount of litigation over edicts from the king's council suggests that the real problem lay with the state's inability to make contracts binding. More radical attempts at judicial centralization like those of Maupeou (the minister for justice) in the 1770s were short-lived. Their only effect was to increase the uncertainty of the judicial process, and they brought no gains for drainage projectors.[32]

Ministers concerned with agricultural development also attempted to subsidize drainage and other agricultural improvements.[33] The subsidies may have been valuable for agricultural improvements that did not involve externalities or public goods problems. For drainage, however, profitability was not the issue. The problem was litigation, and it could not be resolved by the sort of subsidies the reformers had in mind – chiefly tax rebates.[34] Subsidies could not reduce the time needed to resolve property rights conflicts, and since the subsidies became effective only after the project was carried out, their financial impact was minimal. Even as powerful a figure as Bertin, the minister of finance and a staunch supporter of agricultural development, could not use the subsidies to avoid litigation. Bertin spent 150,000 livres on litigation in the 1780s on a marsh that would have been worth a little over a million livres drained. Furthermore, he spent the money over a space of ten years. It is clear that the 150,000 was probably not the total sum he would have spent on legal fees to drain his marsh because his suits were still in court in 1789. The subsidies Bertin was eligible for would have increased his profits had he been able to drain his marsh, but they had no impact on the delays that prevented him from draining it.[35] As Bertin no doubt found out himself, in the absence of a clear rule for compensating customary right, lawsuits were almost inevitable and subsidies worth very little.

A final set of reforms undertaken to raise royal revenue actually increased the number of lawsuits. Liberal interpretations of a point of medieval law known as the *vacance* rule led nearly every community in Normandy into lawsuits.[36] The *vacance* rule was a medieval law that gave all abandoned (*vaine et vague*) land to the king. It had been used by medieval kings to repopulate deserted areas, but in the modern period little land was sufficiently devoid of activity to be legally recognized as abandoned. Projectors could, however, attempt to have a marsh or a fen recognized as abandoned in order to secure their property rights. Indeed, once land had been recognized as abandoned, the king could grant it out again. If a piece of land could be found that was vacant, the subsidies offered to projectors became very interesting, because only a small purchase price would be paid and no taxes were due for the first twenty

---

[32] See Sutherland (1986, 23). Sutherland in fact argues that the judicial reforms brought no gains for anyone, whether projectors (because the reforms failed), judicial officials, or agents of the state (because the prestige of all civil servants was badly damaged).

[33] Edit de Compiegne (AN H¹ 1499, August 13, 1766). See also AN H¹ 1496–7.

[34] See AN H¹ 1487–90.   [35] See Forster (1980, 77).

[36] The suits over the *vacance* rule left a mass of archives in AD Calvados, C 4190, 4203.

years. Several individuals attempted to use this clause to secure marshes that were in fact the common lands of villages. They argued that the lands were not under cultivation, but lack of cultivation did not mean lack of use, for village communities owned strong customary use rights to most marshes. As the *intendant*'s delegate in Caen remarked:

With regard to the *généralité* of Caen there is not one foot of commons, fen or marsh in the hands of the King due to lack of cultivation. These lands are not abandoned, nor deserted. Since time immemorial all the communities enjoy their communal property privately and pasture them each year without interruption.[37]

One of the most extreme attempts to interpret the *vacance* rule liberally was launched in 1761 by M. de Boullonmoranges, a refugee from Turkey who had converted to Catholicism and was the protégé of high court nobles. Under the *vacance* rule, Boullonmoranges was able to secure a royal grant of 12,240 hectares of land. Of that, 1,600 were marshes in more than a hundred parishes in the *généralité* of Caen. The royal grant gave him property to all of the 12,240 hectares unless other individuals could prove ownership. Not surprisingly, the grant gave rise to scores of lawsuits, since all those concerned (parishes and their lords) marshaled evidence on behalf of their title to the land claimed by Boullonmoranges.[38] This was litigation on a scale never previously experienced. Bures, a parish whose only interest was a small share of 500 hectares of fens and marshes, spent more than 3,000 livres in just four years over the Boullonmoranges affair. And during the same four years, Bures actually paid a total of 4,571 livres defending its rights to the marsh in other suits. During these four years, Bures's expense was equal to the revenue of nearly 30 hectares of arable land, a vast sum considering Bures's small stake in the marsh. The total amount of money expended by Bures against Boullonmoranges is not available because the accounts of the community are incomplete; however, because litigation lasted for more than twenty-two years, it is safe to assume that Bures spent significantly more than the 3,000 livres it went through in the four years for which there are data. Boullonmoranges himself spent an enormous sum on litigation. Precisely how much is difficult to evaluate, but Boullonmoranges suggested that his cost had run to more than 300,000 livres. And one can only guess how much the other parishes in the *généralité* spent.[39]

The trials over the *vacance* rule lasted until 1783, when an edict was published forbidding concessions of the sort given to Boullonmoranges.[40] But for twenty years the validity of property rights over marshes had been so shaken that no one would have dared to drain a marsh. And

[37] AD Calvados, C 4203 (June 1765).   [38] AD Calvados, C 4197–4203.
[39] See AC Bures, 563 E D T/15. For Boullonmoranges expenditures, see AD Calvados, C 4200.
[40] AD Calvados, C 4200 (62).

while by 1783 the litigation over the Boullonmoranges affair had somewhat clarified property rights to marshes, the Revolution intervened before anyone could discover whether that clarification would aid drainage.

The Old Regime reforms had thus failed to reduce litigation costs, and no changes were made in the institutional structures to facilitate private ventures into drainage. So many overlapping property rights over marshes existed that no development could take place without litigation. The reform attempts of the second half of the eighteenth century were thus doomed to failure because they did not resolve the central problems: overlapping property rights and endless litigation. As I shall argue in Chapter 8, the Old Regime monarchy simply could not resolve these central issues. The simplification of the property rights structure would have demanded deep social change, for much of the economic and political power of the church, the nobility, and other elite groups came from seigniorial rights. One might argue that they could have sold their seigniorial rights to individuals, but transaction costs made this process impossible. Alternatively, the state could have bought these seigniorial rights, but the process would have raised insurmountable political obstacles because of the distributional consequences. Only a revolutionary government could have effected institutional change of such magnitude.

The fiasco of the Boullonmoranges episode did not deter the government from promoting drainage. In fact, in the last years of the Old Regime, faced with the failure of private enterprise, *intendants* resorted to using a royal agency to advance reclamation: the Ponts et Chaussées. It drew up plans for the Divette Canal, which improved drainage in about a fifth of the Dives Basin.[41] The Ponts et Chaussées then built the canal and administered it between 1783 and the Revolution. To pay for the cost of the canal, landowners were taxed on the basis of acreage. Communities near the canal sued to reduce their share of the burden, but the amount of litigation was far less than it had been with the Marais des Terriers. Suits were rare because the project demanded no redistribution of land. Litigation was also limited because the projector was the state, and it clearly had a strong hand against plaintiffs. A public agency had thus replaced private enterprise. In nineteenth-century Normandy, even stronger public institutions would supervise the division and drainage of common lands, and administrative decision would replace litigation.

———

The French Revolution and the Napoleonic regime left France with a strong central authority, a reduced uncertainty about property rights, and a strengthened administrative agency (Ponts et Chaussées) capable of handling all the technical aspects of drainage. As a result, drainage

[41] AD Calvados, C 6771, and S 1004a.

flourished, with two types of projects undertaken in the first half of the nineteenth century. First, common lands were drained and sold to local landowners. Second, the Ponts et Chaussées worked to increase the flow of the Dives, for as more and more drainage took place, more and more water had to be moved out to sea. The institutional challenge was to allocate the costs of both types of projects among landowners.

The Revolutionary years (1790–1814) were slack years for drainage despite the fact that the estimated rates of return were higher than either before or after: about 200 percent for Troarn and 80 percent for the Marais des Terriers. During the same period, no drainage occurred, and there is significant evidence that existing networks fell into disrepair. As noted in Chapter 4, the lack of development and upkeep suggests that, despite the marked demand for drainage as evidenced by high rates of return, the institutional structure had failed. The absence of drainage is not surprising, for the period of the French Revolution was a time of extreme turmoil and was propitious to investment. The Napoleonic period that followed replaced the upheaval of internal strife with that of international conflict. The state had little interest in questions like that of drainage; it exerted its newfound authority in other directions. Thus, it was not until after 1815 that drainage projects could benefit from the new institutional structure.

While the period between 1789 and 1815 was one of inactivity in terms of drainage, the Revolution did recast the institutional constraints on the supply of drainage. Foremost among the reforms was the creation of a powerful executive that wielded far more power than even the absolute monarchy. The Revolution, in fact, achieved the centralization of power that the monarchy desired but could never obtain. The new power of the executive meant that its decisions about drainage matters would carry far more weight because there could be little doubt about their legality. At the same time, the judicial system was completely reformed. The Revolution ended the tensions between legal authorities and the executive, and the likelihood of winning a judicial appeal of an administrative decision declined dramatically.[42] The judiciary also lost jurisdiction over a host of economic matters; henceforth, these matters were attributed to *prefets* – the nineteenth-century counterparts to *intendants* – and to the Ministry of the Interior. Such reforms alone would have been enough to reduce the institutional costs of drainage significantly, but the Revolutionary governments also gave greater coherence to village government. Municipal councils became executive and legislative bodies whose decisions were subject to the approval of the *prefet*. Once the *prefet* had approved municipal council resolutions, they could not be opposed by

---

[42] In fact, after 1807 all litigation about common property took an administrative route. See Block (1856, 1208–18).

villagers. Hence, projectors after 1815 found a state that was more co-operative than the Old Regime and able to wield much greater power.[43]

Beyond these institutional reforms, the Revolution also rewrote property rights. All seigniorial property rights were destroyed. Church property was nationalized, and whatever land that was under cultivation – either arable or pasture – was sold to the public. The sale of *biens nationaux* (land that had been confiscated from the church or from nobles who fled France during the Revolution) put large amounts of property on the market and completely redrew the distribution of landownership. Before the Revolution, the Abbey of Troarn had been the largest landowner in the Dives Basin. After the sale of the *biens nationaux,* large secular landowners took its place. The nationalization of church property and the end of seigniorial privileges also gave the state and local communities exclusive ownership of the marshes that had once belonged in part to seigniors. Where customary rights existed, as in the Dives Basin, the municipalities received all the undrained marshland. In effect, the end of feudal *privileges* as well as the sale of church property swept away all overlapping property rights and secured the villages title to the marshes they used for pasture. The Revolution had eliminated one of the major parties to Old Regime litigation – the seigniorial lord. Once he was gone, litigation subsided.

The Revolution also facilitated the division of village commons, including marshes. The consent of every villager, a necessity for almost every project under the Old Regime, was no longer required after 1815. Subject to the approval of the *prefet,* a favorable vote in the municipal council was now enough to divide and drain a common marsh. Projects were subject to the review of both the *prefet* and the Ministry of the Interior, but once the central government agreed to a scheme, individuals had little power to resist drainage. The state could enforce whatever rule it chose for allocating costs and benefits while offering few avenues for judicial appeal by property rights owners. The effect was to eliminate much of the uncertainty that had hung over drainage projects.

The very magnitude of the institutional changes suggests why it took so long for some areas to reap any benefits. Throughout the 1790s, there was widespread resistance to the Revolution and its succeeding regimes. The structure of property rights enacted by the Revolution was anything but secure. Both land and credit markets were very severely affected by the Revolution. Drainage projectors may well have feared that improving land to which property rights were insecure would be an unprofitable operation. Had projectors kept the improved land and leased it, they still

---

[43] See Sutherland (1986, 344–7) and Petot (1958, chap. 4). Petot, among other historians, stresses the centralization and administrative logic achieved by the Revolution and the Napoleonic era.

would have required substantial institutional stability. Both the Terriers and the Troarn projects required eight years to turn a profit. In other words, beyond the time it took to complete the improvement, a projector would have to have been guaranteed at least four years of rental revenues from the Terriers project and nearly seven from the Troarn project for discounted revenues to equal discounted costs.[44] Yet during the period 1789–1815, there were few years in which individuals could be reasonably confident that the government would withstand such a test of time. New governments might invalidate the Revolution's redistribution. Thus, it was not until 1815, when the Restoration monarchs affirmed the institutional changes of the Revolution, that drainage could proceed.[45]

Under the Restoration, only one problem remained: finding a set of rules that would promote drainage and preserve the new structure of property rights. It is clear that *prefets* and ministers alike saw the division of the commons not only as an economic task − a means of increasing the cultivated acreage − but also as a political one − a means of increasing the number of grateful and conservative landowners. Furthermore, just as the *intendant* had controlled the budgets of communities, so *prefets* controlled the budgets of municipalities, and they did so with political goals in mind. They exercised the same control over the *syndicats,* the associations of landowners that administered drainage projects.[46] One key difference between the *syndicat* and its Old Regime predecessor was that the decisions reached by the *syndicat* could not be appealed once authorized by the *prefet.*[47] This applied in particular to decisions about allocating costs, and it allowed the *syndicats* to divide up the costs and benefits of a project without the risk of paralyzing opposition.

Most laws governing the division of common lands and the creation of *syndicats* were enacted between 1800 and 1823. They had a substantial impact in Normandy. Between 1820 and 1848, most villages in the Dives Basin secured royal authorization to divide and drain their com-

---

[44] The Terriers project still required at least eight years of institutional stability because it required four years to drain.

[45] Hence, the central difference between the enclosure movement in England during the French wars and improvements in France was not relative prices − food was scarce in both countries over this period. It was not the text of the law; improvements were favored in France and enclosures were promoted in Britain. Rather, Britain was institutionally stable and France was not.

[46] If the projects were very large, they required the approval of the Ministry of Interior. See AD Calvados, S 1269, for the role of the Ponts et Chaussées and the municipalities in the administration of the *syndicat*. See also Block (1856, 1087–90, 1208–18).

[47] See AC Janville, 9 E 344/46 (1831–2), for an interesting exchange between a village council and a *prefet* over division rules for the marshes. The *prefet* forced the village council to accept the rules laid down by the central government within a year and without litigation. Moreover, he alone decided who was eligible to receive a share of the drained marsh.

mon marshes. Marshes and other commons were to be sold in equal portions to village households, and the law allowed households to mortgage their portions to the village at 5 percent for twenty years.[48] Drainage of the marshes was accomplished at the time of the division. The engineers of the Ponts et Chaussées drew up plans and the *prefet* made sure they were carried out. A *syndicat* was also created to ensure upkeep of the drainage network.

The creation of a larger, supravillage *syndicat* contributed to the solution of the perennial problems with the Dives exit into the English Channel and with the maintenance of levees on its banks. Created in 1821 by a decision of the central government, the Dives *syndicat* oversaw all drainage in the basin.[49] Each village had at least one representative on the *syndicat*'s council, but the real power lay with the *prefet* and the engineer of the Ponts et Chaussées. The engineer proposed improvements, and the *prefet* approved the budgets. This *syndicat* provided, for the first time, an institution that collected funds from all concerned parties (landowners and communities) for the maintenance of levees throughout the Dives. Although the *syndicat* was unable to raise funding for a general drainage project, it did reduce flooding in the Dives area because levees were now built and maintained by a single agency.[50]

The achievements of the first half of the nineteenth century may well seem limited: a couple thousand hectares' worth of drainage in Calvados and only a few hundred in the Dives. But if we discount the period between 1789 and 1820 as too tormented politically for long-term investments, the new institutional structure was able to achieve in three decades what the eighteenth-century regime had failed to do in a hundred years. Most drainage projects that floundered in litigation before 1789 were undertaken and completed by 1850 before new technologies became available. The governments that succeeded the Old Regime demonstrated remarkable continuity where drainage was concerned. They all promoted drainage, and they did so effectively. Their effectiveness resulted from a new distribution of property rights and a new set of political and judicial institutions, and the institutions that mattered were the product of Revolutionary reform. The success of drainage in Normandy after 1815 highlights the institutional and economic importance of the Revolution of 1789.

---

[48] All the localities in the Dives area that owned some marshland seem to have divided their commons between 1820 and 1848. One cannot be completely sure, however, because of the loss of village archives during the fighting in 1944.

[49] AD Calvados, S 1269, "Reflexions de M. Pfistre-Duvant, ancien president du syndicat" (1829).

[50] AD Calvados, S 1270–1.

Throughout the eighteenth century, administrators and investors had tried to increase agricultural output by promoting drainage. The record shows the extent of their failure: Between 1714 and 1783 no significant drainage projects were realized in the *généralité* of Caen. Yet drainage would have been profitable under a more favorable set of institutions. For example, recourse to an administrative solution of the sort adopted after 1789 would have been more efficient than the existing legal process for determining the validity of property rights claims. Yet while such a solution was repeatedly attempted under the Old Regime, it was never fully implemented. In the face of litigation over customary rights, all the theoretical powers of the *intendant* and council vanished, perhaps because the state felt that drainage did not warrant a confrontation with provincial authorities.

In eighteenth-century Normandy, resources were effectively expended on litigation that did not foster economic growth. Everyone – except the lawyers, of course – was made worse off. The Old Regime government could not resolve the problem, for the government itself could not write binding contracts; hence, there was no certainty in property rights over marshes. The rewriting of property rights and centralization of authority effected by the Revolution offered a potential setting for renewed investment in agricultural investments. That setting was only a potential until the Restoration made the sweeping changes of 1789 definitive.

7

The development of irrigation in Provence, 1700–1860: the French Revolution and economic growth

Water control, as noted in the preceding chapter, involved both removing excess water from fields (drainage) and supplying water to the land when necessary (irrigation).[1] Throughout France the goal of many projectors was to improve land by increasing either drainage or irrigation. In the well-watered North and West, projectors sought to address primarily problems of drainage. In the more arid South and East, however, they concentrated on irrigation as in the region known as Provence, which is the focus of this chapter and which provides another example of the importance of the Revolution's institutional change. After examining the evolution of technology and relative prices in Provence between 1700 and 1855, I argue that the division of political power blocked all attempts to increase the supply of irrigation under the Old Regime.

As one of France's most arid regions, Provence was an area where the development of an irrigation network should have had the greatest impact before 1789.[2] Years when rainfall is negligible from June to October are frequent, restricting agricultural production to grains, grapes, and olives on dry fields. Both before and after the Revolution, it was argued that the obvious remedy to the arid climate was irrigation. An eighteenth-century historian, Joseph Fornery, wrote:

There is in Cavaillon [a small town in Provence] considerable commerce in artichokes, peas, garlic and beautiful fruits. The water of the Durance River is responsible for this rich produce. This water, as we said above, admirably enriches

[1] A slightly different version of this chapter appeared in the *Journal of Economic History* 49 (Sept. 1990): 615–38.
[2] Throughout the chapter, Provence will denote the present-day *départements* of the Vaucluse and the Bouches du Rhône. While these departments represent only lower Provence, the rest of southeastern France, namely the Côte d'Azur and upper Provence, has a much lower potential for irrigation development. For details see Maps 1.1 and 1.2.

the soil and the inhabitants [of Cavaillon] make a great profit from it. . . . If the river itself is dangerous, its water by contrast is excellent. It carries silt so rich that it makes the most meager lands fertile. . . . The canal of Oppede, which distributes water to a good share of the territory of Cavaillon, is responsible for the best produce in the region, and this justifies what I said about the extreme utility of building a great irrigation canal across the province.[3]

In fact, the example of Cavaillon, the town with the most irrigated acreage in Provence, was frequently used to argue for further irrigation development. Fornery's praise of irrigation was echoed by Jean de Villeneuve, the *prefet* of the Bouches du Rhône in the early nineteenth century:

The territory of eighteen communities has been fertilized to such an extent that the value of their land has doubled, a striking proof of the value [of building an irrigation canal]. . . . Under a burning climate where sometimes nine months go by without rain, where the northwest winds blow so frequently, where the limestone or sandy soil is made even dryer by the deforestation of the mountains, irrigation is a necessity. This is something one cannot repeat enough: all the efforts of farmers tend toward irrigation and it should be the goal of every improvement in Provence.[4]

Until the twentieth century, the main source of irrigation water was the Durance River, a tributary to the Rhone. Because the water of the Durance is very silty, it acts as a natural fertilizer, which permitted eighteenth-century farmers to avoid the biennial fallow on irrigated plots. The abandonment of the fallow alone indicates how dramatic an impact irrigation could have on total output. Yet the value of output would have probably more than doubled because irrigation also allowed farmers to abandon traditional crops in favor of fodder grasses, peas, beans, and other high-value crops. These more valuable crops require both the warmth of the summer and a significant amount of water. Thus, irrigation could lead to substantial per acre increases in output.

A more accurate measure of the increase in efficiency associated with irrigation is the estimated rise in total factor productivity, a measure of productivity change that takes into account the fact that more labor and capital were applied to irrigated land than to dry land.[5] Using sharecropping contracts to trace changes in the quantity of labor and capital applied to the land, I estimate the total factor productivity per

[3] Fornery (1903, Vol. 3, 501–2). Fornery's words echoed those of the regional assembly when a new canal was proposed. AD Vaucluse, C 34, fl. 478.
[4] Villeneuve (1825–9, Vol. 3, 714).
[5] Replacing the fallow with cultivation leads, over two years, to twice the output on the same piece of land but at the cost of more labor and capital. Farmers probably invested some labor and capital on the fallow so that irrigation would not double labor and capital inputs. Because I want to compute a lower bound for total factor productivity growth, I assume that labor and capital inputs double.

acre would have risen at least 30 to 40 percent as a result of irrigation.[6] Thus, irrigation would have represented a significant increase in efficiency.

Unlike farmers or landowners, Old Regime royal governments may have been less concerned with increasing production on specific plots of land than with raising regional agricultural output. Using data from the 1870s, I compute a conservative estimate of the change in total factor productivity as a result of the development of irrigation after 1789 (see Appendix 2 for details). Had the canals planned or proposed under the Old Regime, but realized only after 1789, been built in the eighteenth century, the increase in total output in the region would have been more than 7 percent.[7] While an output increase of 7 percent at the regional level may seem small, it would have significantly eased any short-term Malthusian constraints on the population, the very problem that concerned so many government officials.[8]

A qualitative survey of the geography and economy of eighteenth-century Provence thus suggests that irrigation should have seen greater development under the Old Regime. Moreover, the benefits of irrigation were well known long before 1820, when the development of irrigation began in earnest. Indeed, some of the canals of southeastern France dated back to the Middle Ages.[9] Many irrigation projects were proposed between 1700 and 1789, so we must look to something other than ignorance as an explanation of why irrigation grew so slowly before 1789.

Improvements in technology, credit markets, and relative price changes are all potential causes for the sudden development of irrigation canals after 1820. Using data from projects built between 1760 and 1860, we shall see that neither profits nor techniques, nor even credit, were the determining factor in the timing of irrigation development.

Let us first assume that there was little technical change and that credit was easily available and then discuss how the data relevant to the issue of relative prices and profits were collected.[10] Too few canals were built

---

[6] Total factor productivity increases of 0.3% per annum were large by early modern standards, so irrigation was equivalent to a century of rapid total factor productivity growth. Cf. Hoffman (forthcoming).
[7] Provence was a net importer of grain throughout the eighteenth century. Thus, the increased output could have been either consumed locally or used to purchase more food. Pillorget (1975, chap. 4, pt. D).
[8] Not surprisingly, Old Regime government officials promoted irrigation, but royal government protection proved insufficient to overcome institutional obstacles.
[9] The canals of Saint-Julien in Cavaillon and l'Hôpital in Avignon were built between 1200 and 1350.
[10] I defend these assumptions later.

The development of irrigation in Provence

between 1700 and 1860 to measure directly the profitability of irrigation projects. Moreover, no irrigation projects were realized until 1765 in Provence. It is possible, however, to estimate the profits that projects would have earned had they been started earlier than 1765. Estimating rates of return requires three kinds of data: price series for the inputs and outputs of canal construction, factor shares for each canal, and an interest rate series (because the costs and benefits are spread over time).

Unfortunately, canal accounts are not very detailed. They do contain enough information for us to calculate factor shares for skilled and unskilled labor, but it is difficult to derive factor shares for such things as quarried stone, lime, wood, and other material inputs. However, since nonlabor inputs were used mostly on bridges and in a few buildings, land and labor composed nearly all the costs of canal construction.

The costs of a canal include building costs and maintenance costs. For the jth construction year, expenditures are divided between man-days spent digging by unskilled labor (d_u^i), man-days of construction, which required skilled labor (d_s^i), and the amount of land consumed by the canal (m). Man-days of labor and acres of land are bought at market prices w_u^t, w_s^t, and p_i^t (all land bought for the canal is assumed irrigated, which biases the rates of return downward). The present value of future maintenance costs is the yearly maintenance cost m_c divided by the interest rate r^t.

The social return to building a canal is taken to be the increase in the price of land when it becomes irrigated, $(p_i^t - p_d^t)$. Since land is the only input in fixed supply, in the long run the net increase in output from irrigation should accrue to the owner of the land. Thus if n acres of land become irrigated, the social return will be n times $(p_i^t - p_d^t)$. Given these assumptions, one can compute the hypothetical benefit–cost ratio R^t had the project started in year t and taken T years to complete: [11]

$$R^t = \frac{n(p_i^t - p_d^t)/(1 + r^t)^T}{\sum_{j=0}^{T-1}(d_u^i w_u^t + d_s^i w_s^t + mp_i^t)/(1 + r^t)^j + (m_c/r^t(1 + r^t)^T)}$$

To estimate hypothetical profits, I was able to construct two different wage series using data from Avignon, a large town in the middle of the area.[12] The data consist of wage bills for unfed labor from the account books of religious and municipal organizations.[13] Religious institutions

[11] The hypothetical internal rate of return is simply the r'^* that set R^t to 1.

[12] For any year, wages are nearly identical across sources in the area. Avignon, the major city, has the most abundant sources and the ones that were used to construct the series. The noticeable intraregional pattern was that unskilled labor was somewhat cheaper in nearby villages, but skilled labor was more expensive there than in Avignon. There was considerable seasonal fluctuation in wages, due partially to variation in the working day. For further details, see Appendix 1.

[13] Most of the workers who received food were paid not on a per diem basis, but on

owned medieval canals and hired labor by the day for maintenance work. Thus, the wage data come from the very professions involved in canal construction and maintenance. These data have been sorted into two series: skilled and unskilled workers (see Table A1.3). The first, unskilled labor, was constructed from the wages of laborers, road gangs, and levee maintenance workers. The second was constructed from the wages of skilled workers (masons, miners, carpenters, gang bosses, etc.).[14]

The data for the land price series are from a sample of land-sale and land-lease contracts negotiated between 1700 and 1855 that were taken from the archives of *notaires* in Cavaillon, the town with the largest amount of irrigated land in southwestern France in both the eighteenth and nineteenth centuries.[15] The choice of Cavaillon allows us to ignore any local market effect on the price of improved land. If anything, the fact that Cavaillon had more irrigated land than other areas should bias the price of irrigated land downward and thus underestimate potential canal revenues.[16]

Calculating hypothetical profits also requires data on the costs and revenues of canals built between 1700 and 1860. Data are available for two eighteenth-century projects, Cabedan-Neuf and Crillon, and for two projects proposed in the eighteenth century but not realized until the nineteenth, Plan-Oriental and Carpentras. While there are insufficient data to estimate the profits of other projects, my sample of canals is representative of most canals built between 1700 and 1860, in terms of size, location, and timing. The canal of Carpentras is as large as any in Provence, and the smaller ones of Cabedan-Neuf, Crillon, and Plan-Oriental are similar in size to most other projects.[17] Table 7.1 displays in con-

---

a monthly or yearly basis. Not knowing either how many days of work corresponded to a year's wages or the value of in-kind compensation, I did not use wage bills of workers who received food as part of their compensation.

[14] The data also reflect some of the extraordinary levels of inflation associated with the French Revolution, unlike most series previously published. For details, see Appendix 1.

[15] Cavaillon is a local market town located seventeen miles to the east of Avignon, thirty miles northwest of Aix, on the banks of the Durance River. See Map 7.1.

[16] In Old Regime France, transportation costs were high. If only a small portion of a given area was irrigated, such land would command a very high price. When the irrigation network was completed, the price would fall dramatically; thus, we want to use a price for irrigated land that is close to the price irrigated land would have commanded after the network was completed and a price for dry land that was the price of irrigable dry land. By 1700, 15% of the area of Cavaillon was irrigated. This large area suggests that most irrigation-specific goods would have commanded only a competitive price. Moreover, most of Cavaillon's nonirrigated land under cultivation was irrigable, so we are in fact measuring the price difference between irrigable and irrigated land with reasonable confidence. See Appendix 1 for details.

[17] Cabedan-Neuf irrigated 600 hectares in and around Cavaillon and was built from

densed form all the project-specific data used in the construction of hypothetical profit streams.

The only other data necessary are interest rates. These are taken for the eighteenth century from *rentes* data collected in Provence, which have been sorted into decadal averages, and for the nineteenth century from French government bond data.[18]

I estimated both benefit–cost ratios and internal rates of return.[19] All projects would have been profitable during nearly the entire period under study. But they would have been more profitable before 1750, prior to being carried out, than after 1820, when they were. Although some projects always appear more profitable than others, changes in estimated profit rates are similar for all projects. In addition, the profitability of an irrigation canal does not seem to depend on the scale of the project. Profits for any project vary significantly from one estimate to the next, but the benefit–cost ratios are less than 1.2 in fewer than 20 percent of the years. The estimates suggest that any uncertainty about the profits of canal construction concerned their magnitude rather than their existence. The dispersion of both rates of return and benefit–cost ratios is due largely to variations in the increase in the value of land as a result of irrigation (75 percent of the variance of the benefit–cost ratios is explained by a regression of the ratios on land prices). Because the hypothetical profits of Old Regime projects are similar to the hypothetical profits realized by projects built after 1820, it is unlikely that changes in technology played a major role in irrigation development. Had there been much technological change, later projects should have been much more profitable.

As Table 7.2 suggests, the highest profits came in the early eighteenth century, between 1700 and 1730. During the years from 1735 to 1755, projects were less profitable – though not unprofitable – than any other time except for the Revolutionary period. A number of projects built after 1760 were proposed during this intermediate period, suggesting that investors, at least, found it profitable to attempt irrigation development. The last decades of the Old Regime between 1760 and 1785 show high internal rates of return and high benefit–cost ratios. The rates of the late eighteenth century were in fact higher on average than those of the nineteenth century, when most of the development actually took place (Fig-

1764 to 1766; Crillon irrigated 1,000 hectares around Avignon and was completed in 1777. Plan-Oriental, another canal in Cavaillon, watered 800 hectares to the north of Cavaillon; it was built in 1823. Carpentras was very large; built in the 1850s, it irrigated more than 4,500 hectares.

18 The interest rate data for the eighteenth century come from Jean-Laurent Rosenthal, "A Credit Market in Old-Regime France, l'Isle-sur-Sorgues, 1650–1788" (UCLA, Dept. of Economics Working Paper, 1990). For the nineteenth century, I have relied on Homor (1977, 156–7, 172, 195–6, 222–3).

19 The data and results are presented in detail in Appendix 2.

Table 7.1. *Canal costs*

Canal	Date of completion	Land irrigated[a] (hectares)	Total building costs[b] (francs)	Capitalized maintenance costs (francs)
Cabedan-Neuf I[c]	1767	500	822,300	97,200
Cabedan-Neuf II[c]	1767	270	172,490	97,200
Crillon	1779	1,000	400,000	400,000
Plan-Oriental	1821	590	138,595	100,000
Carpentras	1857	5,000	5,297,011	100,000

Canal	Years under construction	Skilled labor (man-days per year)	Unskilled labor (man-days per year)	Land requirements (in hectares)
Cabedan-Neuf I[c]	2	88,815	73,053	27.0
Cabedan-Neuf II[c]	2	21,410	19,824	27.0
Crillon	3	35,088	41,190	49.5
Plan-Oriental	2	63,561	46,631	9.7
Carpentras[d]	6			
First three years	—	61,341	224,242	96.7
Last three years	—	221,852	112,403	73.3

[a] One hectare equals 2.4 acres.

[b] All costs are given for the year in which they were incurred. There was no need to deflate them because they are converted into quantities of labor.

[c] Cabedan-Neuf I and Cabedan-Neuf II are the same canal, but the various sources on construction accounts could not be reconciled. Not knowing which one was more accurate, I present results based on both sets of sources.

[d] Carpentras was a very large canal. For the first three years, work focused on the main canal. Only in the next three years were branches built. See Caillet (1925, Vol. 1, 69–70).

Sources: For the first estimate of the costs of Cabedan-Neuf, Barral (1876, 539–44). For the second estimate of the costs of Cabedan-Neuf, Syndicat du Canal de Cabedan-Neuf (1883, 45–52). For the canal of Crillon, Reboulet (1914, 37–40) and Barral (1876, 326–7). For Plan-Oriental, Martel (1955, 394–5) and Barral (1876, 545–7). For Carpentras, Caillet (1925, Vol. 2, 199–201) and Barral (1876, 325–6).

ures 7.1 and 7.2). After 1785, the rates of return were highly erratic until 1820, no doubt because of the uncertainties provoked by the Revolution.

The high estimated profits through most of the Old Regime, and in particular during the years from 1700 to 1730, suggest that changes in relative prices were not responsible for the late development of irrigation in southeastern France. During most of the eighteenth century, rates of return were in fact higher than they were in the nineteenth century. Yet irrigation development was much more limited from 1700 to 1789 than

Table 7.2. *Average hypothetical internal rates of return for irrigation projects (percentage per year)*

	Period				
Canal	1700–30	1735–55	1760–85	1790–1820	1820–55
Cabedan-Neuf I (1767)	113.0	32.5	77.8	11.0	63.3
Cabedan-Neuf II (1767)	60.0	2.1	35.3	−16.0	24.2
Crillon (1779)	91.8	33.8	68.6	10.7	57.8
Plan-Oriental (1821)	126.9	49.9	104.0	25.3	78.9
Carpentras (1857)	32.0	13.4	30.0	−1.4	25.0
Interest rate	5.0	5.1	5.0	7.5	4.3

Source: Table A1.4.

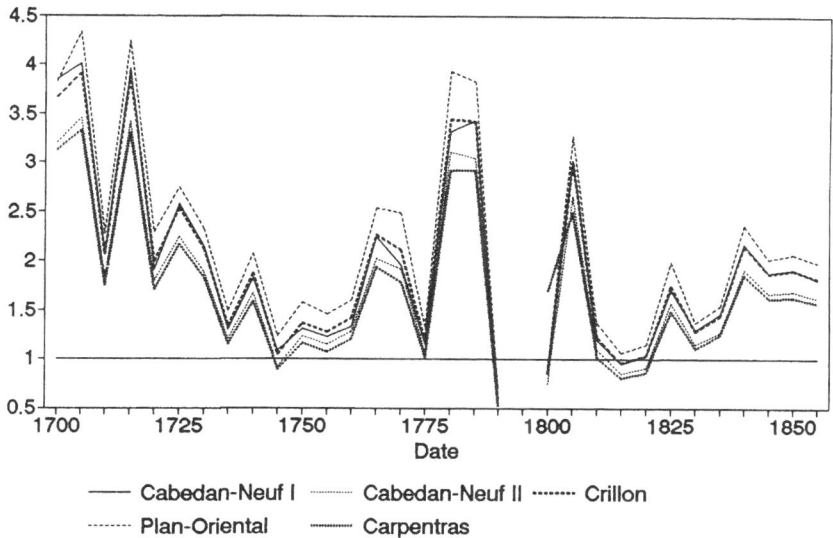

Figure 7.1. Hypothetical benefit–cost ratios for Provençal irrigation projects. *Sources:* See Appendix 2.

it was from 1820 and 1860. Indeed, despite the fact that hypothetical rates of return before 1760 were well above the interest rate, no canal was built before that date. Thus, some sort of a market failure in the supply of irrigation must have been at work in the eighteenth century.

Given the high hypothetical profits, it is important to examine technology and the availability of credit. If we were to look at transportation canals, we could see signs of technical change. By 1750, transportation

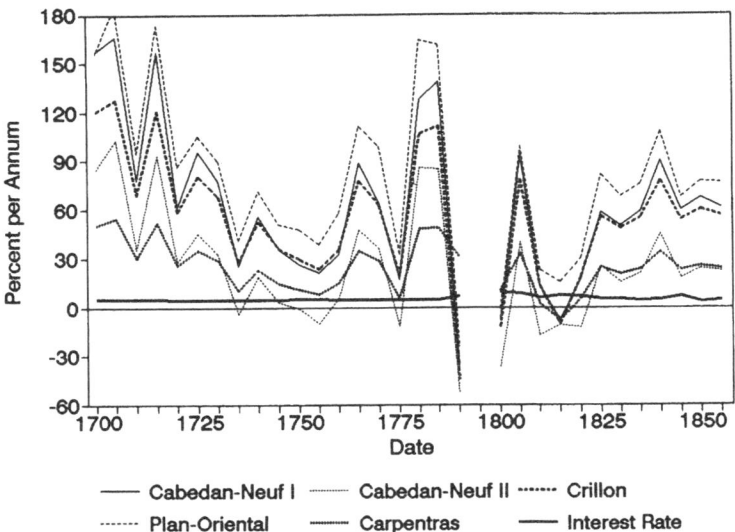

Figure 7.2. Hypothetical rates of return for Provençal irrigation projects. *Sources:* See Appendix 2.

canals had become very sophisticated. They involved locks, dams, bridges, and complex water management.[20] The technology of urban water supply was also greatly improved between 1700 and 1855.

But we are interested in irrigation canals, and their technology seems to have remained the same from the Middle Ages to the late nineteenth century. The methods used between 1700 and 1860 resembled those used in the thirteenth century in building the canals of Saint-Julien and L'Hôpital, or in the sixteenth century in building the canal of Craponne.[21] From 1200 to 1870, all new irrigation canals were unlined dirt ditches, where water flowed by gravity alone. Stone masonry was used only for bridges. The only dams in use, flimsy dirt levees that captured the water from the Durance River, had to be rebuilt after every large flood. They diverted part of the river's flow, but no attempt was made to retain water in a reservoir.

Agricultural development simply could not support the innovative, but very expensive technologies used for transportation canals or for urban water supply. The low degree of technological sophistication utilized in agricultural projects stands in contrast to the higher technology of several urban water projects undertaken in the nineteenth century. One project,

[20] A valuable source on eighteenth-century canal technology is Delalande (1777). See also Maistre (1968, chap. 3).
[21] Rigaud (1934); Caillet (1925, chaps. 2 and 3).

the canal of Marseille (1840–8), a joint urban and rural water supply canal, offers a good example of the technologies available in the nineteenth century but not yet in use in agriculture. The canal of Marseille featured a large dam and a permanent reservoir and many bridges, and it ran underground for 25 percent of its length. The project was financed by the city of Marseille, which attempted to sell excess water to farmers. The city also wanted farmers to pay a share of the building costs equivalent to their share of the water. This made the price of water fifteen times the price on other agricultural projects. As a result, the scheme to retail excess water to farmers failed.[22]

Despite the available technology, the methods used to build irrigation canals did not change. Yet it is possible that the experience gained from past canal construction led to smaller engineering errors. The resulting reduction in risk would have increased the viability of projects by lowering the risk premiums demanded by investors. Yet technological risks – the risks associated with the construction phase of the project – seem to have been very limited. Even in the eighteenth century, the relationship between technology and cost was well established. Engineering costs could be predicted with a good deal of confidence because of the experience gained from building transportation canals, which were much more complex and thus riskier. Irrigation projects were, by contrast, very simple, even when there were unanticipated delays or higher than expected costs.[23] It thus appears that between 1700 and 1860 change in the methods of irrigation canal construction was limited, and technological risk did not threaten irrigation projects or constrain the supply of irrigation.

Because the construction of irrigation canals involved considerable cash outlays, the assumption that credit was easily available is crucial for my argument. While Old Regime France lacked a well-developed, centralized credit market, the limited development of credit markets did not block the expansion of the irrigated area of Provence.[24] In defense of this position, one can marshal four kinds of evidence. First, the credit demands of most irrigation canals were tiny relative to the credit demands of Provençal villages that borrowed extensively during the Old Regime. Second, credit markets based on mortgages, which were very active in rural France, could have provided significant sources of capital for irrigation

---

[22] Masson (1929–30, Vol. 7, 162–7). This canal ran nearly 100 kilometers through a very rugged part of Provence to deliver water to Marseille.

[23] The canal of Boisgelin, the most ambitious canal constructed before the Revolution, shows that risk was small. The engineer, Brun, had warned of the very large costs associated with the tunnel. Yet he did not doubt that the tunnel could be built; BM Méjanes, Ms. 840(853).

[24] Goubert (1973, Vol. 2, chap. 7) offers an introduction to Old Regime finance. See also Chaussinand-Nogaret (1976).

promoters, who were often wealthy landowners.[25] Third, Jewish agents were directly involved in making loans to at least one projector in the 1780s. These loans did not have the collateral of land, and the projector paid an interest rate double that of the mortgage rate – 8 to 10 percent as compared with 5 percent. Fourth, the high nobility was also able to finance projects directly in the case of many of the smaller projects, perhaps putting its vast wealth at the disposal of canal promoters because irrigation canals did not carry the stigma attached to many other forms of investment.[26] Therefore, it seems there were sufficient sources of capital (though not necessarily through organized markets) to carry out irrigation projects.

Finally, the apparent market failure in irrigation cannot be ascribed to a lack of acumen on the part of Old Regime investors. In fact, the magnitude of entrepreneurial activity is striking when it is contrasted with the failures endured by canal promoters before 1789. Every canal built after the Revolution can be traced back to a serious promoter under the Old Regime who had expended considerable resources attempting to secure all the authorizations needed to build the canal. These promoters failed overwhelmingly, if we measure success by the ability to build a canal and earn a profit. The failure rate remains very high even if we stipulate only that a canal was built. Indeed, the pre-Revolutionary expansion of the irrigated area represents only 16 percent of what was actually planned before 1789 and built before 1860. Simple economic arguments do not seem to explain the failure of irrigation development under the Old Regime or its success after the Revolution. Instead, it seems that the peculiar fragmentation of power that characterized the Old Regime constrained Provençal canal promoters.

While the failure of irrigation supply had multiple causes, the most important, I would argue, lay in the division of authority over rights of eminent domain.[27] This problem was well understood before 1789, yet

---

[25] The canal of the Midi was financed primarily by the estates of Languedoc through loans. See Forster (1960, 66–74), Beik (1985, 292–7), and Maistre (1968, chap. 4).

[26] See Masson (1901, 423–5), Elie (1953, 112–13), and Reboulet (1914, 46–7). In the case of the canal of Crillon, 25% of the construction costs were advanced by Jews and another 25% by nobles and bourgeois. Landowners were the largest source of credit in France because they could borrow money through mortgages. Had promoters been able to interest more than a small number of landowners, the credit problem would never have existed.

[27] Other causes of failure were the costs associated with securing water rights and the severe revenue problems related to the fact that most of the costs of the network were sunk when the builder bargained with landowners to sell them water

resolution was elusive because eminent domain authority was embedded in the Old Regime structure of *privileges*.

The woes of a sixteenth-century canal builder illustrate the costs of divided authority. In 1554 Adam de Craponne, a Provençal nobleman and engineer, received a royal grant to draw water from the Durance. In order to secure eminent domain rights for his canal, Craponne had his grant acknowledged by the local assembly – the Estates. Yet some Provençal communities (called Terres Adjacentes), did not come under the jurisdiction of the Estates as far as eminent domain was concerned. These villages delayed the project until Craponne gave farmers there unlimited, free access to the canal's water.[28]

Despite these outlandish concessions, Craponne completed his canal in 1559 and sold a number of irrigation rights. In dry years, however, the Terres Adjacentes villages used up most of the canal's capacity, and with no water to deliver Craponne had to renege on his other contracts. The resulting suits led Craponne to early bankruptcy and discouraged other investors from pursuing irrigation projects. From the standpoint of Terres Adjacentes villages, the whole affair was a free ride. Although Craponne's bankruptcy saddled them with part of the maintenance costs, they now received irrigation water without the burden of any construction costs. Divided authority over eminent domain could indeed create severe problems for canal developers.

The structure of authority Craponne encountered in the sixteenth century was a legacy of medieval state building, and it remained in place until the Revolution of 1789. After the division of Provence between the Pope and the counts of Provence in the twelfth century, the Pope's share became known as the Comtat Venaissin (hereafter, the Comtat). The Comtat corresponds to the present-day *département* of the Vaucluse. The counts of Provence retained control of the Comté of Provence and the Terres Adjacentes. The western half of the Comté of Provence (hereafter, the Comté) and the Terres Adjacentes make up what is now the *département* of the Bouches du Rhône.[29] In 1481, the king of France inherited

rights. For details, see Chapter 8. Although these other causes were important, they were due to the same division of authority that encouraged rent seeking over rights of eminent domain. Focusing solely on rights of way simplifies the argument.

[28] On the canal of Craponne, see Bertin and Autier (1904), Rigaud (1934), Villeneuve (1825–9, Vol 3, 698–714), and Masson (1929–30, Vol. 7, 148).

[29] The Terres Adjacentes were a set of administratively independent communities that included Marseille, Arles, and a number of villages on the border between the Comté and Comtat. These communities had never been directly incorporated into Provence. In fact, until they became part of France, the Terres Adjacentes recognized only the direct authority of the Count of Provence. The best reference detailing the political divisions of Provence is Baratier (1969). For more detail, see Masson (1929–30, Vol. 4) and Villeneuve (1825–9, Vol. 3).

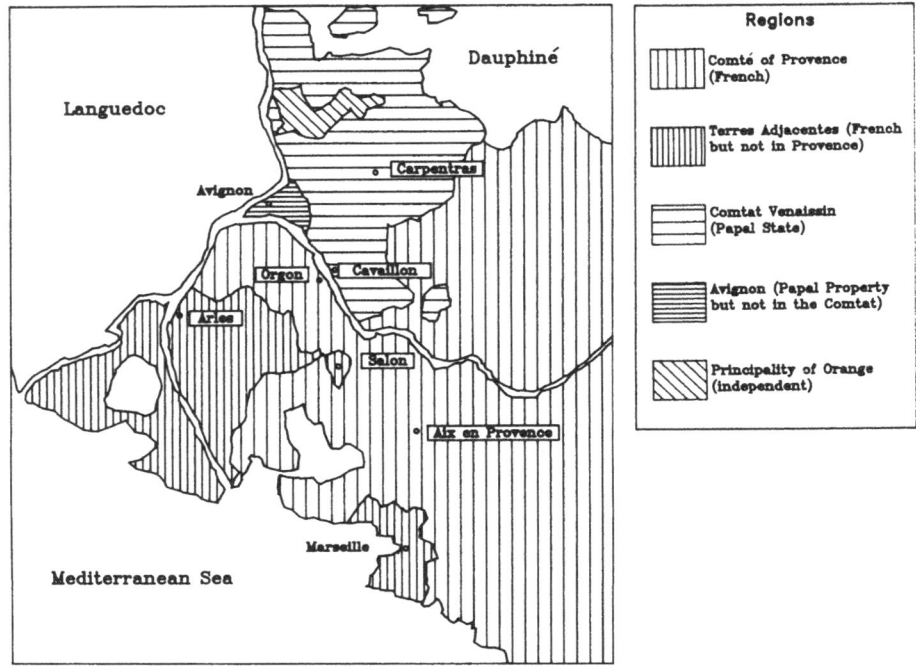

Map 7.1. Provençal regional boundaries in the eighteenth century. *Sources:* Masson (1929, Vol. 4) and Villeneuve (1825, Vol. 3).

the Comté and the Terres Adjacentes. The administrative and political boundaries are displayed in Map 7.1.

The complicated geographic divisions just outlined corresponded to organizational divisions that seem to have determined the transaction costs of irrigation. Before the Revolution, two organizations alone should have decided the fate of irrigation projects in the Comtat, although, as this chapter will make clear, their authority was far more limited. These organizations were the Estates of the Comtat, a representative assembly in charge of taxation, and the Apostolic Chamber, the Comtat's final court of appeals. The approval of the Estates was necessary to secure financial or legal support for irrigation projects, but the Pope and his local representative (the vice-legate) had veto power over decisions by the Estates, a veto power they regularly exercised. Similarly, the Apostolic Chamber was a court of last resort and should have enforced all contracts relating to irrigation. In fact, appeals could be made either to the Apostolic Chamber itself or in some rare cases to papal courts in Rome.[30]

[30] Elie (1953, 112–13); Reboulet (1914, 37–50).

The development of irrigation in Provence

The Comté had organizations similar to those of the Comtat. As a French *pays d'état* it had, like the Comtat, a fiscal and legislative body: the Assemblée du Pays. Like the Estates of the Comtat, the Assemblée du Pays could provide a locus of bargaining for institutional change. As far as the judicial system in the Comté was concerned, the final court of appeals was the Parlement of Aix-en-Provence.[31]

The Terres Adjacentes, the final area under study, were classified as a *pays d'election*. These communities were directly under the authority of the French king and had no estate. In these villages, the division of judicial authority among the king, the villages, and the Parlement of Aix was very ambiguous. Most important for this study, individual villages rather than a central authority seem to have controlled eminent domain rights. In the Middle Ages, the Terres Adjacentes had been autonomous and had in fact decided issues of eminent domain alone. Under the Old Regime, the extent of local autonomy was uncertain and subject to erosion by the Crown. Yet the Terres Adjacentes villages were well organized and could credibly threaten to sue anyone who did not secure rights of eminent domain from them.

One might assume that the political border that ran between the Comtat and the Comté was the root cause of the institutional problems, but in fact the two territories were divided by the Durance River. Thus, most canals were either in the Comtat or in the French part of Provence, even though nearly all drew water from the Durance. So most Comtat affairs were strictly Comtat affairs, and the same was true in the French part of Provence. Moreover, the problems of eminent domain were sufficiently important within each political division that we can ignore the effect of the Comtat's independence from France. Let us, for example, consider rights of eminent domain in the Comté. Since any canal on the southern side of the river would irrigate land mostly in the Comté, the king, the Estates, and the Parlement would all be involved in granting rights of way; however, the need to cross the Terres Adjacentes added a further cost even though it, too, was part of France. In the Comté, the best sites to draw water from the Durance were in, or led into, the Terres Adjacentes. Thus, villages that ruled over eminent domain in the Terres Adjacentes could block or delay projects.[32]

It is hard to test the hypothesis that divided authority over rights of way made it very difficult to build irrigation canals. Nonetheless, it is possible to examine the history of five Old Regime canals to see whether institutions significantly raised the costs of irrigation. One relationship emerges from these histories: The more that canals crossed institutional boundaries, the more difficult they were to build.

Of the four small canals completed under the Old Regime, three were

---

[31] Masson (1929–30, Vol. 7).   [32] Bertin and Autier (1904, 113).

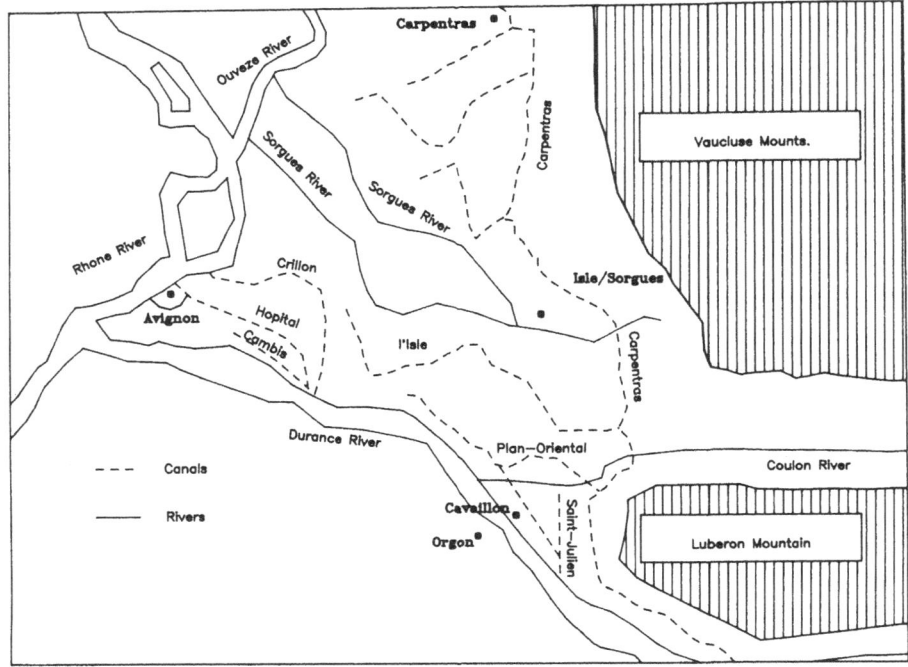

Map 7.2. Irrigation canals in the Vaucluse. *Sources:* Barral (1876), Masson (1929, Vol. 4), and Villeneuve (1825, Vol. 3).

in the Comtat and the fourth was in the Comté. Each of the four projects distributed water to, at most, a few communities. They did not cross any important political boundaries, yet even among the four projects delays and transaction costs rose with size. All the canals appear in Maps 7.2 and 7.3.

The two smallest canals, Janson and Cambis, were each only a few kilometers long and faced only minor transaction costs. Each was financed entirely by the principal landowner – the marquess of Janson and the duke of Cambis, respectively – who wanted to irrigate his very large estate. Both the marquess and the duke maintained strong political ties to the French royal court, and they successfully lobbied for water grants. Their large estates eliminated the free-rider problem and allowed each nobleman to internalize most of the benefits of his irrigation canal. In his grant application, the marquess of Janson argued that the benefits to his estates would more than suffice to cover the construction costs.[33] He did

[33] AN H¹ 1515 (March 1780).

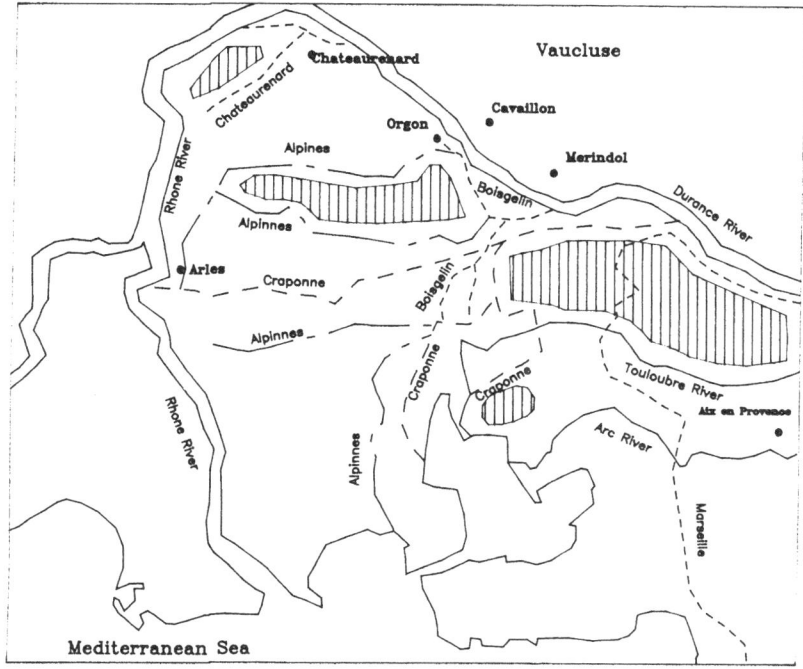

Map 7.3. Irrigation canals in the Bouches du Rhône. *Sources:* Barral (1875), Masson (1929, Vol. 4), and Villeneuve (1825, Vol. 3).

allow the neighboring community to use the canal for irrigation purposes, no doubt to facilitate his use of rights of eminent domain. But there is no evidence he or Cambis failed to make a profit from the canals, even though the villages did not contribute to construction costs. In any case, unlike the other examples, these two small canals were completed swiftly and had few transaction costs.

The third project actually completed was the canal of Cabedan-Neuf, built in the Comtat around 1765. Although it affected only three communities, Cavaillon, Les Taillades, and Merindol, it was large enough to create problems with eminent domain. The canal was built by an association of landowners under the tutelage of the city of Cavaillon. Because most of the land irrigated by the canal was either in the territory of Cavaillon or in that of Les Taillades, the costs of the canal were apportioned between the two villages according to the area irrigated. The third village, Merindol, enjoyed a generous free ride. Much of the canal passed through Merindol, which, unlike Cavaillon, lay in the Comté, not the Comtat, and thus was not subject to the powers of eminent domain of

Comtat authorities. Not surprisingly, Merindol sued Cavaillon over rights of eminent domain. The issue was settled out of court: Merindol received water from the canal, but it did not contribute anything to the project.

Except for the redistributive implications, the free riding by Merindol was relatively unimportant: It did not stop the project. Litigation was avoided because Cavaillon alone could have paid for the entire canal and still benefited from the project.[34] Yet the history of Cabedan-Neuf demonstrates that the involvement of a mere three communities was enough to drive institutional costs higher than when only one community was involved. These institutional costs were associated with scale because of the extreme division of authority in the region.[35]

The fourth canal, the canal of Crillon, delivered irrigation water to Avignon and surrounding communities. It was built by the duke of Crillon, descendant of an old line of Comtat noblemen who had led the French king's armies. Using his favor at court, Crillon secured a grant to draw water from the Durance. He then had the grant registered in the Parlement of Aix. Next he secured rights of way from the city of Avignon that were recognized by the Estates of the Comtat and by the vice-legate. The canal, however, ran through several communities and challenged the water monopolies of a number of seigniors and monasteries, all of whom held the project up for ransom by attacking it in court. The most important suit was brought by the duke of Gadagne, lord of Vedene, one of the communities traversed by the canal. Gadagne contested Crillon's right both to cross into Vedene and to cross Gadagne's irrigation canal. The suit was temporarily settled out of court in 1777, and in the settlement Gadagne granted rights of way in return for water rights. The settlement was not fully executed by either party, and the case was still being litigated after the French Revolution.[36]

The canal of Crillon demonstrates the need for precise geographic and historical detail. Gadagne could litigate against Crillon only because the canal's rights of eminent domain had been first granted by the city of Avignon and only then approved by the Estates. Avignon and the Estates had a complex relationship because the city was in fact a Terre Adjacente of the Comtat, having been bought by the Pope from the counts of Provence in 1348. As a result, the authority of the Estates over Avignon was

---

[34] Syndicat du Canal de Cabedan-Neuf (1883, 48–69). Cavaillon chose to bargain with Merindol directly rather than with the Assemblée for a right of eminent domain. Presumably, both Merindol and the Assemblée were seeking rents, and Merindol proved cheaper to pay off.

[35] Across the Durance, in the Comté, the town of Châteaurenard also attempted to build a canal in the 1780s. Châteaurenard was also forced to negotiate over rights of way and water rights with the nearby town of Noves and its seigniors. See Barral (1875–6, Vol. 1, 375–6).

[36] BM Cecano, Ms 2549. Appeals were heard in the Apostolic Chamber and then in Rome throughout the 1780s. Again, the settlement gave free water to Gadagne.

unclear. Although the Estates and other Comtat organizations had approved the canal, they had not specifically granted rights of way in the Comtat. Thus, the validity of the duke of Crillon's rights of eminent domain was subject to dispute and formed an open avenue for anyone to attack the project.

The history of the canal of Boisgelin, my fifth example, shows the costs of fragmented authority in a large-scale project, built in the Comté under the financial authority of the Assemblée du Pays after a number of other attempts had failed. The proposed canal had two possible routes: one ran through the Comté alone; the other crossed the Terres Adjacentes. While the latter would have been cheaper, it involved bargaining with the Terres Adjacentes for rights of eminent domain. Rather than bargain with each village in the Terres Adjacentes, the Assemblée du Pays avoided the issue but paid a very high price.[37] The Assemblée opted for the all-Comté route — a much more expensive alternative from an engineering standpoint because it involved tunneling through about one kilometer of solid rock near the village of Orgon. The cost of tunneling totaled nearly 400,000 livres and absorbed half the yearly budget of the canal for eight years. Yet piercing the rock of Orgon allowed the promoters to avoid the Terres Adjacentes villages of Sénas and Salon, where the cheaper route lay. Once the tunnel was built, the Assemblée had the ability to exclude the Terres Adjacentes from the benefits of the new canal if they did not contribute to its cost. Not surprisingly, the Terres Adjacentes communities purchased a significant amount of water from the canal just before the French Revolution, and a branch canal through Sénas and Salon was built.

Unlike all other irrigation projects, which involved little more than the digging of ditches, the canal of Boisgelin had to resort to an extraordinarily costly technology, a technology imposed by institutional constraints. Once again, the division of authority led to much higher transaction or institutional costs than if only a small canal had been built. In this case, these institutional costs took the indirect form of digging a tunnel at Orgon rather than bargaining or litigation.

Thus, the histories of a few projects make it clear that the institutional environment blocked irrigation by raising the cost of canal building. The obstacles had their origins in the long-term development of institutions in southeastern France. One had either to pay off obstructionist villages, as Craponne did, or to bear much higher construction costs, as did the promoters of the canal through Orgon.[38] The presence of organizations like the Estates and the Parlement did allow for some institutional change. It was, after all, possible to build the canal of Boisgelin. But the sort of

---

[37] See Villeneuve (1825–9, Vol. 3, 714–21).
[38] BM Cecano, Ms 16050–2459, 4°6189; Reboulet (1953, 41–4).

institutional change that would have substantially reduced costs lay outside the authority of these organizations. In fact, neither the king, the Parlement, nor the Assemblée could reform the Terres Adjacentes. Their peculiar status indeed constituted a *privilege,* something only the Revolution would change.[39]

Irrigation was easy prey for rent-seeking villages because it involved both economies of scale and significant geographic specificity. Canals were networks; hence, the costs involved in building the main canals did not rise as quickly as the irrigated area increased. Moreover, because canals relied on gravity to move water, each area usually faced a single economical drawing site from the river. As a result, villages close to the Durance could credibly threaten irrigation projects with much higher costs or insurmountable engineering problems if they refused to grant rights of eminent domain. Most often, villages were in a position of such strength that promoters could only give in or give up.

The phenomenon of villages holding irrigation projects up for ransom was not due to the specific form of village organization in eighteenth-century Provence. In fact, as the well-known examples of the sale of judicial offices and the monopolies of craft guilds suggest, rent seeking was commonplace under the Old Regime.[40] Ironically, in the case of irrigation the greatest rent seeker of them all, the Crown, was generally allied with canal promoters against local powers that were holding the projects up. Yet the Crown proved powerless to resolve the problem in the case of irrigation.

––––––––

For twenty-five years after 1789, there was no increase in irrigated area in Provence, and those networks already in use were very poorly maintained.[41] Revolutionary turmoil during the years from 1789 to 1795 was violent in Provence. Moreover, starting in 1792, warfare drained away manpower and drove up the price of labor relative to land, a problem that grew even worse during the Napoleonic period (1798–1815). Yet even though the Revolution caused delays in the extension of the irrigation network, it was bringing about institutional reforms that would set the stage for future development, notably the construction of a number of new irrigation canals after 1820.

Institutional reforms, initiated by Revolutionary regimes and contin-

[39] The Terres Adjacentes took advantage of Provence for much more than irrigation. See Villeneuve (1825–9, Vol. 3, 755–61).
[40] See, e.g., Mousnier (1971) and Bossenga (1988, 405–26).
[41] AD Vaucluse S (Usines et Cours d'Eaux, Cavaillon and L'Isle sur Sorgues). The series S was being classified and sorted at the time I looked through it; thus, no precise references can be given.

ued by Napoleon, would drastically cut the institutional costs of irrigation in the nineteenth century, consolidating all powers of eminent domain in the hands of the central government and destroying the old organizations and institutions that had prevented reform. In Provence, the annexation of the Comtat and the abolition of the peculiar status of the Terres Adjacentes removed two major obstacles to development of irrigation. For the first time since the early Middle Ages, a single authority could decide all issues of property rights in Provence. Beyond the simplification of regional boundaries, the most important Revolutionary reform was the centralization of legal and political power. Although centralization had been one of the goals of the absolutist monarchy, and although the king had held veto power over virtually all economic activity, he had never been able to eliminate local organizations like the *parlements,* the Assemblée du Pays, the Estates, or even village councils. Centralization during the Revolution eliminated these local organizations and replaced them with a single pyramidic administrative structure headed by the Ministry of the Interior. In the case of rights of way, the agent of the government at the local level – the *prefet* – was now charged with making all decisions.[42] The destruction of all other veto players freed irrigation development from the shackles of strategic behavior. Towns and villages near rivers could no longer refuse rights of way for new irrigation projects simply to protect the market value of their older irrigated land or, even worse, to siphon off part of the profits.

Revolutionary reforms gave *prefets* complete authority over projects until they were built and removed the judiciary from the planning stages of irrigation, making it difficult for local groups to delay projects through litigation. Local groups could only appeal a project before the *prefet,* whose approval was thus sufficient to guarantee the success of an irrigation project. Litigation – when it occurred – did not start until after the canal was built and the social gains were realized. Moreover, conflicts over technical and engineering issues could no longer be litigated but were decided by French administrators. After the Revolution, not only did the central administration have the power to provide promoters of irrigation with the property rights they needed, it also had the power to enforce all the contracts.[43]

After the end of the Napoleonic regime in 1815 and under many different governments, irrigation in southeastern France flourished. There was considerable state help, including engineering advice, administrative oversight, and the full power of its newly centralized authority. One form

---

[42] Petot (1958, 383–7), Bergeron (1972, 33), Sutherland (1985, 345), and AD Vaucluse S, Usines et Cours d'Eaux.

[43] Ponteil (1965, 30–4).

of support, however, was conspicuously absent: The government offered very few subsidies for the development of irrigation.[44] By and large, the irrigation canals of the nineteenth century seem to have been paid for by the landowners whose fields were irrigated, further evidence that institutions rather than technology or profits had caused the earlier market failure.

Whether in the case of a small project such as the canal of Plan-Oriental, or in the case of a large project such as the canal of Carpentras, state approval was decisive. The Plan-Oriental canal involved only a small amount of land (800 hectares) and delivered water to fields in only a few villages. The project was quickly approved by the *prefet* and completed in 1823, less than four years after initiation. In contrast, the canal of Carpentras involved more than 4,000 hectares in many different communities. Although the size of the canal slowed development, the state showed the flexibility of its new power by designing organizations with authority over many communities and many canals. For example, an organization was created that legally grouped all the canals drawing water from the Durance at the site originally used by Cabedan-Neuf alone, thereby allowing an efficient sharing of this desirable site.[45] Because the promoters were able to rely on the support and authority of the national government, the canal of Carpentras was completed in 1865, less than twenty years from its launching.

The overall success of irrigation in the nineteenth century is striking: More than 16,000 hectares, at least half of all the land irrigated from the Durance by 1875, received water from canals completed between 1820 and 1860. In all, more than 80 percent of the increase in irrigated area between 1700 and 1860 came after 1820.

Under the old Regime, the division of authority over rights of eminent domain limited the scale of irrigation development. In Provence, the political division of authority – a legacy of the Middle Ages – gave ample opportunities to a variety of groups to hold up projects. Villages successfully used this position to extract rents from canal promoters.[46]

Only local irrigation projects could avoid the costs associated with divided authority over rights of eminent domain. As a result, the transaction costs associated with irrigation development increased dramatically when projects crossed authority boundaries. Irrigation promoters were forced to face these transaction costs because the state proved incapable of reform. The problems of eminent domain were simply not

[44] After 1870, however, irrigation development almost always involved government subsidies.
[45] Caillet (1925, 75–6).
[46] Veto power was widely used to extract rents from developers in Old Regime Provence; see Baehrel (1962, 450–6), Pillorget (1975, 196–207), and Agulhon (1970, 43–59).

resolved before the French Revolution, which makes it surprising that any irrigation projects were developed before 1789. By contrast, the nineteenth century witnessed substantial growth in irrigation in south-eastern France without significant litigation and with much shorter delays than had been customary in the previous century. Between 1820 and 1865, the area irrigated in Provence more than doubled, and all the water in the Durance came to be used. Hence, as far as irrigation is concerned, it appears that 1789 had a dramatic effect on transaction costs.

PART III

Property rights and litigation under absolutism

8

The weaknesses of monopoly power

Under the Old Regime, drainage and irrigation were stymied by litigation over property rights that dragged on for decades. Most of the courts, like the *parlements* or the *conseil* in which suits over water control were argued, were supposed to be final courts of appeal. Despite repeated claims of final and expeditive justice, however, the courts continued to hear appeals; hence, it is clear that they reneged on their promises of speedy trials. The courts did not keep their promises because they were not bound to honor their announcements.

As we shall see, kings, courts, and private individuals frequently announced future plans of action and later chose not to follow those plans. The inability of many actors in pre-Revolutionary France to commit to future plans dramatically increased transaction costs in some sectors of the economy. Solutions to problems of inconsistent announcements, although well known before 1789, were never effectively implemented under the Old Regime.[1] In contrast, the governments that followed the Revolution steadfastly abided by contracts. More important, after 1800 the state forced other organizations to keep their promises as well. This chapter presents an investigation of the problems of commitment under the Old Regime in three areas that proved crucial to the development of water control: litigation, royal granting policies, and water rights.

To uncover the logic behind the actions of Old Regime organizations, let us examine the case of judicial officers in greater detail. Since civil litigation was quite costly in monetary terms, justice should have been prompt and appeals should have been discouraged unless they had good cause. Yet because justice was venal, judges had a pecuniary incentive to hear appeals, and as a result they almost invariably heard them.[2] Moreover, judicial officers taxed trial costs, which meant that their revenues

---

[1] There were repeated calls for judicial reform and much interest in British constitutional government. See Sutherland (1986, 22–3).

[2] The same pecuniary interest also explains the difficulty of out-of-court settlements. Indeed, appeals against out-of-court settlements would be heard, and some would also be invalidated.

depended on the effort that litigants put into the trial. Because first trials were unlikely to settle cases, litigants invested fewer resources early on than if the *parlements* had been definitive jurisdictions. Thus, the returns to first trials were lower for the justices than if they had behaved as a final court of appeals.[3] However, judicial officers reaped benefits in the long run because they were able to tax the same cases repeatedly as they went through numerous appeals.

We can make sense of the judges' behavior if we think about the incentives they faced. During the first trial, *parlement* judges would presumably have wanted litigants to behave as though litigation were definitive and to invest a great deal of effort in the trial. To that effect, justices announced forcefully that *parlement* verdicts were final. Yet when faced with an appeal, *parlement* officers frequently heard it because of the revenue it would bring. Judges wanted litigants to believe their announcement that litigation was definitive when they preferred to hear appeals. Despite the fact that the only consistent announcement would have been for the officers to say that they would hear appeals, they announced that verdicts were definitive.

The problem faced by the *parlement* officers was that verdicts were durable services. Once a case had been decided definitively, litigants no longer needed the court's services. When, however, judges overturned their own decisions, litigants continued to require the services of the court. Hence, judges wanted litigants to believe they were producing a durable verdict, but once having produced the durable good they had an incentive to destroy it.

Problems of durability have been modeled in an extensive theoretical literature.[4] The results have recently been applied to problems in economic history, in pioneering articles by Douglass North and Barry Weingast and by Avner Greif.[5] Most of the theoretical literature focuses on durable goods sold by a monopolist, following a seminal paper by Coase.[6] Although that literature accurately describes the problem faced by canal projectors who attempted to sell off irrigation rights, it does not directly illuminate other Old Regime problems of durability. Moreover, the economics literature has not examined in detail how such problems can be resolved.[7] Hence, it is best to build a model that captures the crucial

---

[3] The problem of repeat litigation was quite acute under the Old Regime. The Crown tried several times to reform the judicial system to no avail. *Parlement* officers attempted to reduce the level of litigation by creating specific rules to allocate cases among competing jurisdictions. Mousnier (1979, Vol. 2, pt 4).

[4] See Coase (1972). More recent work in the field includes Bulow (1982), Kahn (1986), Gul, Sonnenschein, and Wilson (1986), and Stokey (1982).

[5] North and Weingast (1989); Greif (1989).   [6] Coase (1972).

[7] More precisely, the solutions considered by the economics literature were implemented neither before nor after 1789.

elements of all the cases and allows us to analyze what alternatives were available to solve these problems.

———

The problem of inconsistent announcements can be described in a two-period model with three participants: a seller and two buyers. Let δ be the discount rate. In the first period, the seller offers to sell either a strong (S) or a limited property right (L) to the first buyer at prices of p_s and p_l, and he must also announce whether he will invalidate or uphold strong property rights in the future. The buyer will earn revenues from the use of the property in both periods: R if it is a strong or r if it is a limited property right. Given the seller's announcement, the first buyer chooses which property right to buy.

In the second period, the seller faces a new buyer who will earn a return of r_2 with a property right that he buys from the seller at p_2. If the seller does not sell to the second buyer, the seller in effect upholds strong property rights. In this case, if the first buyer bought a strong property right in the first period, he now earns R in the second period. If the seller contracts with the second buyer, the seller invalidates all strong property rights so the first buyer only receives r in the second period.

In most durability models, prices are endogenous; in the problems we want to examine, that was not always the case. For example, the state oversaw the cost of the judicial procedure; that is, it put a limit on how much judges could receive from litigants. We will thus take prices as being exogenously given and only assume that $p_s \geq p_l \geq p_2$. Without loss of generality we can normalize the profits to limited property rights to zero, $(1 + \delta)r - p_l = 0$, and those of new buyers to zero as well, $r_2 - p_2 = 0$.

Figure 8.1 shows the model in extensive form. The strategic process has three moves: First, the seller announces whether he will uphold or invalidate strong property rights; second, the first buyer chooses a property right (strong or limited); third, if a strong property right has been sold the seller decides whether or not to invalidate it. To complete the foundations of our analysis, it is necessary to rank outcomes explicitly according to what would maximize aggregate income. Since the game involves only transfers from buyers to the seller, we can assume that social welfare is the sum of the buyers' returns. Let the social welfare function be $W(x, y)$, where x is 1 if a strong property right was sold in the first period and x is 0 if a limited property right was sold. Similarly, y is 1 if a second-period sale occurred and 0 otherwise. The social welfare function can take on three values: $W(S, 0) = (1 + \delta)R$, $W(S, 1) = R + \delta(r + r_2)$, and $W(L, 1) = r + \delta(r + r_2)$. Social rankings are introduced to consider which institutional arrangements are socially efficient and under what circumstances institutional reforms may be warranted. It is also

Property rights and litigation under absolutism

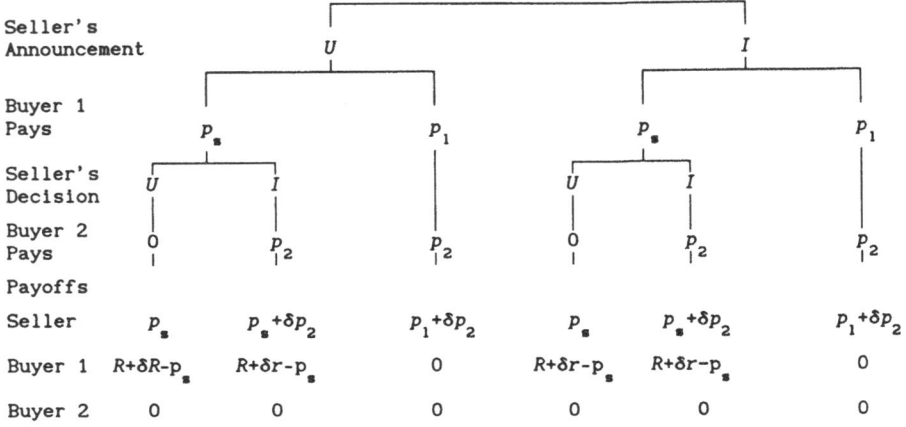

Figure 8.1. The durability game in extensive form.

important to uncover the relationship between initial conditions (values of δ, r, R, r_2, p_s p_1, p_2) and the reforms that may be necessary to achieve social improvements.

I focus on solutions to the problem such that individuals maximize their income at every point in time, taking the behavior of others as given – in economists' words, only subgame-perfect Nash equilibria will be examined.[8] Only such solutions are examined because most of the participants in property rights exchanges under the Old Regime had a good deal of experience with the political and judicial system. Thus, they would have been unlikely to make decisions that were irrational or that did not take the future into account. A solution to the problem is defined by a triplet (A, B, C), where A is a first-period announcement by the seller (uphold or invalidate strong property rights), B is a first-period decision by a buyer of what property to buy (strong or limited), and C is a second-period decision by the seller about strong property rights (uphold or invalidate). Let us analyze the process under two different institutional structures, depending on whether the seller can commit to his announcement. The results are summarized in the following propositions and proofs. In Proposition 1, the seller cannot commit to his announcement. In Proposition 2, the seller can make his announcements binding.

---

[8] If fact, the requirement for a subgame-perfect equilibrium is stronger than those given in the text. The requirement is that strategies be optimal at every point in time, even for events that never occur in equilibrium. Thus, subgame perfection is in essence a restriction on threats that can be part of a Nash equilibrium. See Kreps (1990). Given the simplicity of the problem, it is unnecessary to discuss strategies explicitly and define equilibria rigorously.

The weaknesses of monopoly power

Proposition 1. In the absence of commitment, if first-period revenues to a strong contract are small enough $(R + \delta r - p_s < 0)$, then there are two equilibria, (U, L, I) and (I, L, I). Both solutions feature a limited sale in the first period and a second sale in the second period. Conversely, if the first period is long $(R + \delta r - p_s > 0)$, then again there are two equilibria, (U, S, I) and (I, S, I). In equilibrium, the seller first sells a strong property right and then invalidates it. In either case, the two equilibria differ only by the seller's announcement.

Proof. At the beginning of the second period, the seller strictly prefers to invalidate strong property rights. Whether a strong or limited property right has been sold, and independent of the announcement, it is still in the seller's best interest to accept the second buyer's offer. In all cases, the first buyer anticipates a second-period invalidation. If $R + \delta r - p_s < 0$, to avoid the loss associated with buying a strong property right the first buyer purchases only a limited one. Hence, the sequence of action is always a weak property right sale and then a second sale independent of the announcement.

If $R + \delta r - p_s > 0$, the first buyer always buys a strong property right in the first period because he makes a profit. The seller sells to the second buyer in the second period. Here again, announcements are irrelevant.□

In order to characterize fully the solutions to the process in the absence of commitment, it is necessary to distinguish which announcements are likely to prevail given that strong property rights will be systematically invalidated. In other words, will the seller tell the truth? To decide which announcements are likely to prevail, let us assume that, from the point of view of the seller, there exists some uncertainty about the first buyer's identity. We can assume that there is a possibility that the first buyer is naive; he makes his decision strictly on the basis of the seller's announcement.

Corollary. In the absence of commitment, the seller never tells the truth. In technical terms, only the equilibria in which the seller announces that first-period sales will be upheld are sequential.

Proof. Assume that there is a probability $\sigma > 0$ that buyers will believe the seller's announcement because the buyers are either gullible or inexperienced. Now assume $R + \delta r - p_s < 0$. If the seller announces (I), gullible players will buy only limited property rights. But if the seller announces (U), gullible players will buy such rights in the first period. In the second period, the seller will invalidate strong property rights. Thus, the seller will make a higher profit by announcing that he will uphold rather than invalidate strong property rights. In fact, for any proportion σ of gullible players, the seller will strictly prefer to announce (U) rather than (I).

Table 8.1. *Equilibria in the durability game*

Initial conditions		Equilibria		Social optimum (case)
		No commitment	Commitment	
$R + \delta r - p_s < 0$	$p_1 + \delta p_2 < p_s$	U, L, I	U, S, U	U, S, U (1) U, L, I (2)
	$p_1 + \delta p_2 > p_s$	U, L, I	U, L, I	U, S, U (3) U, L, I (4)
$R + \delta r - p_s > 0$		U, S, I	U, S, I	U, S, U (5) U, L, I (6)

Note: Equilibria are defined as triplets (A, B, C), where A is the seller's announcement, B the buyer's purchase decision, and C the seller's second-period decisions. U, Seller upholds strong property rights; I, seller invalidates strong property rights; L, first buyer buys a limited property right; S, first buyer buys a strong property right.

The same argument will also work in the case where $R + \delta r - p_s > 0$ if one assumes that some proportion σ of first buyers will buy only strong property rights if the seller announces (U). □

Traditionally, the literature on time consistency has focused on the seller's desire to announce credibly that he will uphold strong property rights because the seller prefers the sale of a single strong property right rather than two sequential sales $(p_1 + \delta p_2 < p_s)$. We want to consider a broader set of durability problems, including those in which the seller has no incentive to protect strong property rights. Thus, we will maintain that a durability problem exists, that the seller would prefer to sell a strong property right and then invalidate it. We will not assume, however, that he prefers a single strong property rights sale over making two limited property rights sales.

Proposition 2. Commitment has only a limited impact on what solutions will prevail. Indeed, if limited property rights have a low price $(p_1 + \delta p_2 < p_s)$ and first-period revenues are small $(R + \delta r - p_s < 0)$, then the seller will be willing to commit himself to upholding property rights. In this case, a commitment to uphold property rights will increase the seller's profits over the no-commitment equilibrium. In all other cases, the ability to commit to announcements will bring no change to the equilibrium.

The proof of this proposition is obvious.

Table 8.1 summarizes the relationship between initial conditions, equilibria, and the social optimum. Cases (1), (3), and (6) are of particular interest for our analysis of Old Regime property rights. As we shall see, case (3) describes the problem of repeat litigation. Case (6) is closest to the problem of royal water-grant policies. Case (1) describes the problem of irrigation right sales and is the closest to the classic durability problem as discussed by Coase and Bulow.[9] It is interesting that commitment could only improve the outcome of one of the three cases we will examine. In the other two cases, (3) and (6), some form of exterior solution would be required to solve the durability problem. In other words, institutional change, like that which occurred under the Revolution, would be necessary to overcome many durability problems

While there are a host of possible institutional reforms to alleviate time consistency problems, I focus on two that were considered under the Old Regime and implemented during the Revolution. The first possible innovation was to separate the individual who benefited from invalidating strong property rights and the individual who made the decision to invalidate strong property rights – this change was at the center of the reform of the judiciary and of national water-granting policies. A second innovation was to have a third party enforce the seller's announcement – this change was implemented in the case of irrigation rights. These institutional changes correspond directly to major themes of eighteenth-century political thought: the creation of a strong executive and independent branches of government. Indeed Old Regime reformers argued that a strong executive would have the power to monitor private arrangements, while independent branches of government would prevent the state from breaking its promises.[10]

Let us return to Old Regime France and examine three problems of time consistency involving the judiciary, the state, and private projectors in detail. The model just analyzed will help us to uncover why these problems proved so difficult to remedy before 1789 and why the Revolution's reforms were crucial to solving them.

The first problem that can be analyzed using the model is repeated and lengthy litigation in Old Regime France. The problem was very prevalent in civil disputes. It extended from the lowest levels of the royal justice system, through the *parlements* to the *conseil d'état*, which reviewed royal legislation over economic matters. Appeals and delays in litigation led to very expensive trials and to unallocated property rights over valuable assets for long periods of time.[11] To underscore the importance of this

[9] Coase (1972); Bulow (1982). [10] Baker (1975).
[11] See Chapters 4, 6, and 7.

131

problem for drainage and irrigation, let us examine a few cases of litigation over property rights in the eighteenth century.

In cases of marsh drainage, as noted in Chapter 4, the central problem was the allocation of land between villages and lords. Both lords and villages based their claims on medieval contracts, while the interpretation of these contracts was left to the judiciary. In the 1680s, the lords of Bourgoin, in central France, attempted to drain the marsh there. Their first attempt failed because of the opposition of village communities. Villagers fought their lords by litigating over the allocation of land among the village, the lord, and the projector. The lords abandoned the project because litigation dragged on without result. In 1766, after highly publicized royal reforms in favor of drainage, the lords of Bourgoin made a second attempt. The suit that resulted was still before the courts in 1789. The case could not be resolved despite the lack of any new evidence because the losing party to any verdict could immediately appeal and receive a staying order while the case was heard yet again. The correspondence of the lords of Bourgoin suggests that they were well aware of the repeat litigation problem, for they appealed to the king to alter the procedure for settling disputes about the allocation of land.[12] Such reforms, although attempted, were ineffective until after the Revolution.

Another case of repeat and lengthy litigation involved water rights in Provence. In the early eighteenth century, when the expansion of the irrigation network was considered, projectors faced the nearly insurmountable problem of finding a site to draw water from the Durance River. Projectors had to fight powerful local magnates who claimed ownership of all drawing sites and based their claims on medieval grants. These ancient property rights tended to be perpetual grants of a site rather than of a limited amount of water, and the magnates argued that hereditary and transferable title to water rights had been granted to their medieval ancestors by the king's predecessors. Thus, they claimed monopoly rights over all water drawn from their site.

In the eighteenth century, projectors with new royal grants to water from the Durance were challenged by the owners of medieval property rights in the court.[13] The litigants were arguing about a purely redistributive question. Had the medieval property rights been enforced, presumably the projectors would have bought water from the owners of these ancient rights. If, however, they had been overturned, development would have proceeded with a loss to the owners of the medieval rights. In any case, the important issue was speedy justice, but that ran directly counter to the interests of the judiciary.

---

[12] An F 10 208.
[13] AC Cavaillon, BB 20–1. See also Syndicat du canal de Cabedan-Neuf (1883, 7–61) and Caillet (1925, Vol. 2).

The weaknesses of monopoly power

Despite the best efforts of the judiciary to delay resolution, one potential projector – the Oppede family – had the stamina to pursue litigation long enough to get a definitive verdict.[14] Litigation over water rights pitted the family against the local bishop and the city of Cavaillon from the early seventeenth century to the middle of the eighteenth. The dispute began when the Oppede family attempted to build a canal through the Comtat using a new royal grant to draw water from the Durance. The grant allowed the Oppedes to draw water from the site that was owned by the bishop; thus, it conflicted with the bishop's monopoly property rights. Indeed, the bishop had received his sole right to draw water from the Durance as early as 1189. To prevent the canal from being built, the bishop sued the Oppede family, demanding that their grant be invalidated.

The conflict resulted in an extremely long suit between the Oppede family and the bishop. No doubt unable to bear the high costs of litigation alone, the bishop sold his monopoly water right to the city of Cavaillon. As a result of the city's entry into the legal conflict, property rights to all water drawn from the Durance in the territory of Cavaillon were subject to judicial review.[15] The process of judicial resolution was complicated by conflicts over which court had authority to review the case, and decades passed without a solution. Finally, after more than thirty years of court battles, the issue was finally resolved by arbitration in 1733.

Such examples of lengthy and repeat litigation abound in areas other than marshland or water rights.[16] Water rights were particularly prone to litigation because royal regulation demanded a judicial determination of all marsh property. In addition, water control was a fixed investment

---

[14] The Oppede family more than others straddled the border between Comté and Comtat. They had become one of the most powerful families of the region. Their power was a recompense for their staunch support of the king in the civil wars of the seventeenth century. The Oppede family owned considerable property in the region. They were ardent speculators in water rights. The first water grant to the family dated to the early sixteenth century, and it had been given to a family member by the Pope for services as ambassador to Venice. In the mid-sixteenth century, they became royal servants and purchased an office of *président* (chief justice) at the Parlement of Aix. See Pillorget (1975, 759–70, 784–90, 856–62) and Caillet (1925, Vol. 1, 30–66).

[15] The city lay claim to a mill and canal network that had been sold by the bishop to the Oppedes in the seventeenth century.

[16] An investigation of trials due to debts in Normandy reveals that, even in local courts, judges were willing to hear appeals of verdicts very frequently. The collection of small debts frequently dragged on for several years to the extent that court costs rose to a substantial proportion of the debt. AC Calvados 1 B, Caen, court costs for the years 1736, 1746, 1756, 1766, 1776, 1786; 5 B, Vire, court costs for the years 1748, 1758, 1768, 1778, 1788; 15 B Condé sur Noireau, court costs for the years 1744, 1754, 1774, 1784. A detailed analysis of this data will be presented in a future study.

and thus required very secure property rights. Reaching an out-of-court settlement was nearly impossible, since even that was likely to be invalidated by the courts. Both in the case of the marsh at Bourgoin and in that of water rights in Cavaillon, the judiciary failed to provide a resolution of the problem. At Bourgoin the first attempt to drain the marsh was simply abandoned, and the second succeeded only after the Revolution changed all property rights. In Cavaillon, resolution was available only through a different method, arbitration, that bypassed traditional organizations. In fact, repeat and lengthy litigation was so prevalent in water rights that the king attempted to replace the judicial system with an administrative procedure headed by the *intendant* – a nonvenal royal official. The effort failed, however, because Parlement officers predictably fought to preserve their judicial authority.

An incident in the history of Toulon, a city in southeastern France, illustrates the *intendant*'s lack of power. In the seventeenth century, Toulon built a canal to increase its water supply. The canal collected water in the village of Ardenne. The inhabitants and the lord of that village were allowed to use some of the canal's water to irrigate their land. At the time of construction, the canal was much larger than the city required. Thus, the villagers had free access to water. Yet as the city developed, water became scarce and conflicts arose. One report on such a conflict noted: "To avoid the costs and slowness of trials that could occur with irrigators, [Toulon] requested that all disagreements over the canal be decided by the *intendant*. This attribution was given by act of the Council in May 1699 . . . and renewed in October of 1748."[17] The attribution of judicial authority to the *intendant* was, however, insufficient to prevent the intervention of the Parlement of Aix when in 1779 the lord of Ardenne diverted more water than he was entitled to by contract. Toulon's municipal officers sued before the *intendant*, but the Parlement, at the lord's invitation, invalidated the *intendant*'s attribution, and in 1782 the officers of Toulon were forced to request yet another royal edict. They argued:

Toulon is a fort. . . . It is therefore important that its mills and fountains run uninterrupted. Yet interruptions would not fail to occur if the daily infractions of owners of land on the banks of the city canals were brought to ordinary justice. The attribution that was given to the *intendant* avoids these inconveniences. It allows him to judge everything summarily, to shorten, depending on the case, the delays of procedure, which in many cases would adversely affect the service of the king. Finally it avoids high expenditure to both sides and it offers them a prompt expedition.[18]

The same argument led royal administrators to stipulate in almost all drainage and irrigation projects that disputes be reviewed uniquely and

[17] AN H¹ 1307 (La Tour, *intendant* in Aix, to Paris: February 25, 1782).   [18] Ibid.

definitively by the *intendant*. Despite the promise of speedy action, however, litigation tended to persist for decades because *parlements* and the *conseil* repeatedly intervened, ignoring the *intendant*'s decisions and supposed authority. Thus, although royal legislation over water promised that the *conseil* or the *intendants* would have a definitive say, all attempts to reduce the length or costs of litigation failed because of the fundamental opposition of judicial officers.[19]

We can use the model described earlier to explain why delays and repeat litigation were rampant in civil disputes in the eighteenth century and why reform proved so elusive. The property right that buyers were attempting to purchase was a verdict. In the case of water rights, verdicts most frequently involved a decision about the ownership of an asset in the form of water or land. Yet verdicts could be invalidated by either being delayed or overturned upon appeal. It seems clear that litigants would have been more willing to spend greater resources both on lawyers and on searching for evidence if trials had resulted in definitive verdicts than if decisions were temporary.[20] Hence, one can differentiate between expenditures for a definitive verdict (strong property right in the model) and expenditures for a temporary verdict (limited property right).

The loser of the first stage of litigation would appeal if it seemed likely that the original verdict would be invalidated on appeal. Thus, losers in the first stage of a trial were analogous to second buyers in the model, while judges, of course, were sellers of verdicts. Because of venality, judges taxed litigation expenditures. As a result, they would have preferred that litigants spend a great deal of resources on each segment of a judicial process. Although judges received more during the first stage of a dispute if litigants believed verdicts were to be speedy or definitive, they still received a substantial amount if litigants believed that verdicts would be overturned. Indeed, a large part of the cost of each stage of litigation was fixed by procedural requirements. In the middle of the process (after the first verdict or in the making of decisions regarding procedural issues that would influence the length of a trial), judges clearly preferred to invalidate verdicts because they would then receive more money.

Had judges preferred speedy over long trials, one might expect that they would have built a reputation for holding speedy trials and hearing few appeals. Since they were organized in corporate bodies, they certainly had the institutions necessary to police themselves had they wanted

---

[19] To be fair, it should be noted that *parlement* officers were protecting their customary privileges of hearing all important cases.

[20] The exact allocation of resources to lawyers and the search for evidence will clearly depend on judicial rules such as discovery and the relationship between the outcome of the first-stage trial and an appeal. Yet expenditures on lawyers should to some extent decline because the arguments in the first trial are less important than in the second.

to do so. Yet in all likelihood, judges derived more revenue from litigation by drawing it out than by curtailing it.

Society, however, draws little benefit from high appeal rates if appeals do not have good cause. For this reason, it would have been socially preferable if most first-period verdicts had been upheld. Thus, case (3) of our model seems to describe accurately the interaction between plaintiffs and judges under the venal system.[21] Unlike the classic case of durability, in which giving the seller the ability to commit to his announcement allows him to maximize his profits in the case of litigation, commitment does not help to resolve the problem of repeat litigation.[22] In fact, the pecuniary advantages of repeat litigation must have played some role in the staunch opposition of Old Regime judicial officers to any move toward reform.

Reform of the judicial process could occur only from the outside. In this sense, attempts to rely on the *intendant* for authority over disputes prefigured the Revolution's reforms destroying venality. Old Regime administrators were aware of the tax imposed on the economy by venal justice. As a matter of fact, protests about the judiciary inserted in the *cahiers de Doléance* focused on delays as well as expenses. Indeed, these delays were also a cost. In many situations, they spelled doom for drainage projects, and certainly other economic activities.

Awareness of the problem was not enough to prompt reform, because the Crown faced a more fundamental problem that could not be dissociated from reform of venality. Venal officers more than anyone appear to have controlled the credit system of the monarchy.[23] Although the Crown attempted to circumvent venal officers by selling life annuities, royal loans on the open market paid high interest rates and were insufficient to quench the Crown's need for cash in emergencies. The necessities of war required raising large sums of capital quickly, and given the state's lack of credibility as a borrower it was forced to rely on intermediaries.[24] In fact, venal officers, because of the large sums of money they had sunk into purchasing their offices, had a direct interest in supporting the Crown. Thus, judicial officers could be relied on for new loans in cases of emergency. In the absence of fiscal reform, there could be no judicial reform.[25]

Moreover, in the absence of judicial reform, it was impossible for *intendants* to assert their authority. The central administration could not

---

[21] The reason for eliminating case (5) is that, since appeals were heard soon after the original trial, R is likely to have been close to r.

[22] Here I use the term "commitment" loosely to suggest either an organizational change or a reputational incentive.

[23] See, e.g., Bien (1987).

[24] The same problem was solved differently in Britain; see Brewer (1988, chap. 1).

[25] See Bien (1987).

be relied on to back up the *intendants* when they entered jurisdictional conflicts with the *parlements*. The *parlements* had to be handled with extreme care because the Crown knew that in the future it might need to rely on the *parlements'* officers to raise cash. All royal reforms, including increased reliance on *intendants*, were at the mercy of the need to raise more revenue.

The Revolution solved the problems of both tax revenue and venality. After 1789, the central government could turn to an assembly to raise taxes and thus was no longer dependent on venal officers to solve short-term credit problems. Venality was abolished so that judges were paid and rewarded by the central administration for administering prompt justice. As if destroying venality was not enough, the Revolutionary reforms led to the establishment of the *cour de cassation,* a court whose only purpose was to review appeals and decide whether they should be heard. The court, however, did not hear appeals. In this system, the court that decided on the frequency of appeals did not derive benefits from appeals, because it could not influence outcomes. Solutions to the problem of commitment in the judiciary were found by changing the structure and incentives of the protagonists in the judicial process. Those solutions, however, were unavailable until the Revolution, when the state ceased to have an interest in the returns to venal officeholding.

Let us now turn to royal granting policies. As we have seen, the judiciary had potent incentives to take full advantage of the litigation that resulted from uncertain property rights. Nonetheless, the king often engaged in schemes that would lead to insecure property rights because it enabled him to raise revenue. The king frequently granted a set of property rights in return for revenue or to offset some past debts. Had the king granted only rights that had not been assigned previously, the process would have yielded both revenue and increased social welfare. New grants, however, frequently conflicted with older grants that the king or his predecessors had made. The new grants were made even though the king had no clear right to renege on the holders of older grants who sued to protect their property rights. The courts were very responsive to such complaints, in part because of venality and in part because *parlement* officers were in effect in charge of protecting *privileges* – those previous agreements between the king and his subjects.

Older royal grants tended to limit economic activity. New grants, by contrast, would have had a positive effect on national income in the long run. For example, the grant to the Oppede family for a new canal would have increased the supply of irrigated fields and thereby would have reduced the price and increased the quantity of irrigated crops. The benefits of new grants, however, were undercut by the expense and delays of judicial contests with the owners of previous grants who faced losses in revenue and prestige.

To illustrate how the king's practice of repeat granting affected property rights in irrigation, let us return to the conflict between the Oppede family and Cavaillon's bishop. The bishop's water rights dated back to 1189, when the count of Toulouse as count of Provence, and thereby owner of the Durance River, had given the bishop a monopoly over the drawing of water from the river. In the thirteenth century, the bishop decided to build the canal Saint-Julien and Cavaillon's first water mill. The bishop gave the city the right to use the canal for irrigation in return for a significant contribution to the building costs. Between 1200 and 1350, the bishop traded most of the canal to Cavaillon in return for a greater participation in maintenance costs by the city. The bishop, however, retained control of his monopoly right to draw water from the Durance. Meanwhile, Cavaillon had rights only to the water in the canal Saint-Julien. Until the seventeenth century, this arrangement suited everyone involved, even the Oppede family, who as the largest landowner in Cavaillon benefited greatly from the canal.[26]

In the late seventeenth century, however, the arrangement began to deteriorate. The Oppede family obtained a royal grant of water that directly conflicted with the bishop's monopoly. The Oppede family and the bishop had had a contentious relationship over water rights during the entire seventeenth century, but as we have seen, the Oppedes' plan to build an irrigation canal brought about litigation at unparalleled levels. The royal courts that handled the suits over water rights were faced with a dilemma. The legitimacy of the bishop's medieval water rights was not really in doubt, since both the bishop and the city owned numerous parchment rolls attesting to the authenticity of the grant. Yet the Oppedes had a royal grant that was equally valid. More generally, if courts had affirmed the medieval grant, it would have been impossible for the king to grant any more water rights and many recent grants would have been invalidated. In short, both sides had received their grants from the king and had solid but conflicting legal rights.

In economic terms, affirming the medieval grants would have allowed local magnates like the bishop of Cavaillon to benefit from development by selling water rights. Conversely, if the old grants were invalidated, the king would have received some returns to further development because he would then have been able to sell more water rights. More generally, the ability to tamper with property rights allowed the king to raise revenue by defaulting on monopoly grants that had been made in the past. After 1600, because the king found it difficult to raise taxes, it became

[26] AC Cavaillon, BB 22 and BB 23.

standard practice for him to alter past property rights in order to raise revenue. Starting under the reign of Louis XIII, every sector of the economy, from judicial offices to water rights, witnessed royal sales of monopoly titles. Soon, however, the king attempted to renege in order to resell parts of monopoly titles to raise further revenue. Because raising revenue remained a central problem for the state until the Revolution, royal defaults on promises were a hallmark of the absolutist monarchy.

In the case of Cavaillon, after nearly half a century of litigation, the bishop's property right was invalidated: It was no longer to be a strict monopoly water right. Thus, while owners of the bishop's water right had unlimited access to unused water in the Durance, they could no longer control access to water. In addition, all the property rights the Oppede family had acquired in the seventeenth century were confirmed. But the court challenges had been costly, not only because they consumed resources but also because property rights to irrigation water were effectively unassigned for four decades. Until the issue of property rights was decided, no development could occur. Clearly, from the perspective of 1700, it would have been better if in 1189 the count of Toulouse had granted a limited rather than a monopoly property right to the bishop. But in the twelfth century, the count of Toulouse was probably more concerned with satisfying the demands of his political ally, the bishop, than with the economic consequences of his action five hundred years hence.

Another case of repeat granting that dramatically affected investment in agriculture concerned marshes. Unlike irrigation, for which litigation and institutional change did bring some improvement, if at tremendous costs, in the case of drainage, problems actually worsened late in the Old Regime. The Crown sought to increase its revenues by offering to grant special privileges to individuals who would invest in agriculture. Most often, these grants assigned property rights to land without any consideration of the fact that the land was already being used. Farmers who used the land protested and litigated, arguing that the king was violating custom, the personal and regional privileges that had resulted from centuries of bargaining between the state and local organizations. As a result of litigation over the validity of new grants, all attempts to drain land ground to a halt.

A blatant example of repeat granting of marshes and low-quality land to a would-be projector occurred in Normandy in 1761. The king granted to M. Boullonmoranges the complete ownership of all abandoned and unclaimed land in Normandy. Boullonmoranges interpreted the grant to imply that he had property rights over all unimproved land in the area. This interpretation led Boullonmoranges to stake claims on almost all marshland in the region, including both village common land and royal land. The state, however, had little authority over either. The king had given up his authority over common land when Normandy was annexed to France. More important, common land was an important source of pastureland for villages and was by no means abandoned. Moreover, almost all royal land had been leased in perpetuity to individuals or villages, so none of it was abandoned either. It appears that the land the king conceded to Boullonmoranges was already owned. Not surprisingly, when Boullonmoranges sought to assert his property rights, villages sued for protection. Litigation over the validity of the grant lasted for more than twenty years and brought all drainage projects in Normandy to a standstill.

The state appears to have underestimated the havoc that was wrought by the Boullonmoranges grant. The benefits of defaulting on the property rights of villages were reaped in the political sphere and may have been substantial. Indeed, Boullonmoranges was protected by two court magnates, the duke of Bouillon and the count of Morange, and it appears that the grant was created to please them. Had the grant been successful, like that to the Oppede family in Provence, the king would have been able to offer significant compensation to the patrons of Boullonmoranges at little direct cost to himself.

The king, or at least his administration, was well aware of the economic consequences of such wholesale infringement on past property rights. In fact, much of the archival evidence on drainage, irrigation, and enclosure comes from files that were put together because of litigation over the redistributive consequences of wholesale property rights reform.[27] One might argue that, had the Crown taken a wider view of the economic impact of its granting policies, it might have refrained from such practices. Yet our model suggests why the Crown may have continued the practice of making grants despite their costs.

During the early phases of negotiation, there was frequently only one demander for a grant of privilege, be it a noble, a guild, a town, or a province. We can think of that lone demander as the first buyer in our model. The king is the seller. In such cases, the king faced a choice of either granting strong or limited privileges – for example, a limited water grant or a monopoly. Strong privileges were worth more than limited ones to recipients, so they were willing to offer more in terms of resources such as loyalty or money. Hence, the king offered strong property rights. These included monopoly grants of water, monopolies for the production of certain goods, or the explicit and permanent relief from some form of taxation.

In this instance, the periods of the model can be thought of as real time intervals of variable length, where the first period was long relative to the second. As a result, R would be very large, and r heavily discounted in the first period. In the second, much later period, another individual or

[27] See Bloch (1929) and Hoffman (1988).

group (our model's second buyer) would be willing to pay for a similar property right. Thus, the king would face an incentive to invalidate the early grant to reap the revenue from the second sale.

The case of Cavaillon illustrates the workings of the early granting process. In the eleventh century, the count of Toulouse, the king of France's predecessor, did not really control Cavaillon, so the only person he could grant water rights to was the bishop. The bishop no doubt preferred a monopoly grant to a simple authorization to draw water because it helped him get the city to participate in construction costs (in the model, $R > r$). Hence, it seems likely that, even though the bishop might have been aware that in the future the king might renege on his monopoly right, that event was sufficiently distant in the future that the bishop preferred a strong property right to a limited one $(R + \delta r - p_s > 0)$. The king for his part liked to sell strong property rights because he was better off that way $(p_s > p_l)$.

The situation remained stable until the count of Toulouse was replaced by a stronger authority, the king of France, and the power of the bishop was sufficiently eroded that others might want to build irrigation canals. More generally, as time passed and the king granted strong property rights, the areas of the economy left untouched by royal revenue collection became scarcer, thereby increasing the incentives for the king to default on his promises and attempt to alter (invalidate) the old grants. Grants could be altered by allowing competitors to produce certain goods for which a monopoly had already been granted, or by raising new taxes, or even by selling more judgeships on the same bench. Undoubtedly, as the king weakened older grants, he alienated the owners of old grants, but he gained the support of the recipients of the new ones. The king also raised revenue. Rewriting grants made by kings every year more distant was less and less a personal breach of promise for the present king. Hence, in the long run, the benefits the king derived from upholding existing grants fell, while the benefits he derived from invalidating them and giving out new grants rose. Thus, at some point the king would find it worthwhile to invalidate old grants.

Had the king been all-powerful, giving out monopoly grants followed by invalidation might have been socially optimal. No one could have fought the invalidation, and such a strategy would have allowed the fullest economic development possible. Yet France's absolutist monarchs were hampered by numerous organizations, and owners of strong property rights systematically fought invalidations in court, a socially costly process. As a result, invalidations did not solely affect the revenues of owners of strong property rights – a redistributive change with little ex post efficiency cost. In addition, invalidations produced litigation, a process that consumed resources; in the absence of invalidations, these resources could have been put to more productive use. Because of the high private and social costs of litigation, first-stage (medieval) grants should have been limited rather than strong. Yet because limited grants led to lower profits for the king than monopoly grants, no endogenous solutions to this durability problem were available. In short, this historical example is identical to case (6) of the model.

Not surprisingly, the Revolution did change the state's incentives. First, reforms eased the constraints on tax revenue and borrowing so that the state found grants an unappealing method for raising revenue. Second, the Revolution eradicated all feudal grants, leaving few old grants standing. In the case of irrigation, the state permanently switched to a regime of granting limited quantities of water so as to reduce the conflicts between holders of old and new rights. Hence, when the state authorized new activities, little litigation resulted.

The final application of the theoretical model does not involve the state except by default. This section investigates why projectors had great difficulty selling irrigation rights under the Old Regime. I ascribe the projectors' failure to the fact that they could not demonstrate that their announced prices for irrigation rights would prevail for any length of time. As a result, landowners did not buy irrigation rights because they were waiting for the price to drop. This led to severe problems for projectors: First, they found it difficult to raise cash for their projects, and second, they found it difficult to convince farmers to sign long-term contracts for irrigation water because of the possibility that the price for these contracts would subsequently fall.[28]

The dilemma of canal projectors stems from the technology of irrigation. Irrigation involves substantial economies of scale and large fixed costs – those that involve building the main canal. So the cost of irrigating an extra acre of land is lower than the average cost of irrigation. The returns to irrigation depend on the quality of the land irrigated and other variables such as distance to market. Thus, some landowners will be willing to pay more for irrigation rights than will others. Let us focus on a simple problem that involves selling irrigation rights in two periods to two types of landowners: those who own high- and those who own low-quality land. This simple problem captures all the important features of the projectors' dilemma and could easily be extended to any number of periods and types of land.

I define as high quality the acreage of land irrigated that would maximize the projector's profits if the market for irrigation rights opened for only one period. I also assume that, if a projector sold irrigation rights to

---

[28] This problem is very close to the original specification of time consistency by Coase and Bulow.

only that much land, he would make a profit. Low-quality land is such that its owners would be able to pay only a low price for irrigation rights; that low price would cover the marginal cost (MC) of irrigation but not the average cost (AC) of irrigation ($AC > p_2 > MC$).[29] Since the projector faces two types of buyers, he can either sell irrigation rights only to the owners of high-quality land at a high price (p_s) or to owners of both high- and low-quality land at a low price (p_2). In the first period, it never pays to charge a low price because, although everyone would buy, the projector would then face a loss because $AC > p_2 > MC$.[30] Hence, the projector always charges a high price in the first period, and farmers with low-quality land will never buy in the first period.

In the context of the model described earlier, since only owners of high-quality land are likely to buy in the first period, they are the first buyers. Owners of low-quality land who never buy in the first period are the second buyers. Projectors are sellers of property rights. To secure credit for a canal, a projector prefers to sell owners of high-quality land contingent contracts committing them to buying irrigation water once the project goes through. In order to do this, he must announce both a price for contingent contracts and a price for irrigation rights after the canal is built. If the announced price is higher (or lower) than the contingent contract price, the projector is in effect announcing that he will uphold (or invalidate) the value of contingent contracts. Buyers can either buy contingent contracts or wait; hence, limited property rights are simply a refusal to buy a water right in the first period ($p_1 = 0$).

If the owners of high-quality land buy contingent contracts in the first period, the projector makes a profit. The problem is that, whenever first-period purchases occur, the projector then wants to lower his price and sell to second buyers in the second period because even at a low price, he makes a profit at the margin ($p_2 > MC$). As a result, first buyers prefer to wait. This will prevent the projector from selling any future contracts for water from the canal.

Since no sales will occur in the first period, the seller will charge a high

---

[29] Clearly there will exist land of many types, but without loss of generality we can focus on only two types of land.

[30] This may seem an unduly strong assumption, yet one can assume that there is much high-quality land and little low-quality land. The problem of time consistency arises as long as it is not in the best interest of the projector to sell water rights at the value of the lowest-quality land. Yet if one ranks all the parcels of land that the canal would reach by the value of irrigation for those pieces of land, some will have low value because they are of low quality or are distant from the market, while others, which are close to town or very fertile, will have high value. Whatever price the projector charges in the first period, if there are first-period sales, the projector will have an incentive to lower his price to capture some of the low-value land. The assumption that selling all land at the low price is unprofitable simplifies the analysis by avoiding the issue of the first-period price.

price in the second period. At this point we must reconsider the model because two cases arise. First, the market may close forever at the end of the second period. Then, first buyers buy irrigation rights at a high price because the projector will have no opportunity to lower his price in the future. Alternatively, one can imagine that the market for irrigation rights will open again for a third period. In this case, the second period is identical to the original first period, and first buyers will again prefer to wait. Indeed, as long as owners of low-quality land do not own irrigation rights, the projector has an incentive to lower his price after owners of high-quality land have bought water rights.

Empirically, since the market for irrigation rights remains open indefinitely after a canal is built, it is not clear what price the projector will be able to charge in order to both sell irrigation rights and make a profit either before or after building the canal. This simple model suggests that, without commitment, only price paths in which no first-period buying occurs are credible.[31] Moreover, unless there is a way for the projector to close the market – that is, to commit himself to no longer selling irrigation rights – second periods become first periods, and no sales in which the projector makes a profit ever take place.

Thus, in the absence of commitment, the projector cannot be assured of a profit. As a result, lenders will not offer credit without collateral, because the expected revenues after the canal is built may not be high enough to cover all the construction costs. Creditors might accept signed ex ante contingent contracts as sources of collateral, yet it proves impossible for projectors to sell any such future contracts. We can therefore conclude that, in the absence of institutional mechanisms forcing commitment, the credit problem is less one of credit markets than a contracting problem between projectors and landowners.

The importance of this theoretical exercise becomes obvious in light of the fact that most canal projectors in the eighteenth century were unable to sell more than a few contingent contracts. Because of these funding problems, many projects never got off the drawing board even if they managed to clear other institutional hurdles.[32] Projectors built irrigation

---

[31] There are other types of contracts that get around part of the durable goods monopoly problem. For example, the projector could offer contracts that promise anyone that his or her price will be the lowest prices he charges at any time. Then in equilibrium, the amount of rights sold is exactly that which would be sold if the market opened for only one period. These contracts, however, would be quite costly to enforce, as each buyer must monitor prices forever. The same monitoring costs are encountered whenever the quantity of irrigation rights that maximizes the projector's one-shot profits is less than the total quantity available, because landowners know that the projector has a long-run incentive to sell everything. Not surprisingly, these contracts were never offered in the eighteenth century.

[32] Floquet, the first promoter of the canal of Boisgelin, failed to interest any local landowners in his project; see Masson (1901).

canals to sell water rights. The larger the canal, the more irrigation rights could be sold and the greater the potential profit for the projector. Building large canals, however, required large amounts of cash. Most often, the sums necessary to build a canal had to be borrowed.[33] Had projectors been able to get farmers to buy irrigation rights before the canal was built, or to commit to buying such rights, the credit problem would have been substantially alleviated. Yet farmers would sign long-term contracts only if they were sure that the price of irrigation rights would not fall once the canal was built.

Projectors could not convince farmers that they would not lower their prices. As noted, earlier projectors had incentives to sell as many irrigation rights as possible, even to the owners of marginal land. Yet those owners would pay less for the irrigation rights than owners of high-quality land. Projectors would have maximized their profits by first charging a high price to owners of high-quality land and then lowering the price for the owners of low-quality land. Owners of high-quality land, however, could foresee the fall in the price of irrigation rights and hence preferred to wait. One might assume that projectors should have discriminated on the basis of land quality – that is, they should have announced different prices for different types of land. Yet such a scheme would require a good deal of information about land types, and a great deal of monitoring to avoid water diversion from low- to high-quality land. In any case, marketing schemes of this sort were never attempted in the eighteenth or nineteenth century.

Other projectors whose personal wealth allowed them to proceed without outside credit never recouped more than a small amount of the construction costs from water sales.[34] In fact, although the total profitability of irrigation was never doubted, it was always difficult to get landowners to pay for projects in the eighteenth and nineteenth centuries. The difficulty was often explained by the lack of foresight and the "backwardness" of peasants. However, those backward peasants may have understood the nature of the process much better than the projectors.

In the absence of commitment, only a company involved in a large

---

[33] No canal built before 1789 seems to have successfully sold irrigation rights even though individuals could accurately measure the value of irrigation. Indeed, the canals of Cambis, Janson, Chateauneuf, and Boisgelin were all built either by public organizations or by private individuals to irrigate their own land. See Chapter 7.

[34] On the canal of Crillon, sales of water rights were insufficient to cover maintenance costs until several years after the canal was completed. At no point before the Revolution were sufficient rights sold to recover a substantial proportion of the projector's investment (BM Cecano, Ms 4°6824). In the cases of Jason and Cambis, the builders gave neighboring villages water at no cost, but they owned enough land themselves to recoup the building costs.

number of irrigation projects would have had the incentives and the ability to build a reputation for not lowering water rights prices. Yet canals were systematically promoted and built by individuals and groups who had interest in only one canal. These one-shot projectors could not solve the commitment problem internally. As a result, under the Old Regime most canals were built by public organizations that did not need to recoup their investments directly.

Only when the Revolution strengthened the central government were solutions to the commitment problem within reach. After 1789, the central government through the *prefets* kept a close watch on irrigation companies and altered their legal status. While some private companies were allowed to remain in operation, most canals built both before and after 1789 became in effect public utilities. To solve problems of administration and to induce farmers to buy future contracts for irrigation water, the state used the flexible organizational form of *associations*. Most important, *associations* solved all the price commitment problems that had plagued irrigation because they were placed under the authority of the state. Thus, an *association* could post a price for irrigation water before the project occurred, and the state would enforce that price forever. The state was the only party to these agreements that could credibly claim to enforce any price path. Although the state could not be bound to obey contracts, it was a repeat player in irrigation development, as it participated in all the projects. Repeated participation gave the state reputational incentives to enforce prices. In fact, the state alone had reputational incentives to enforce announced price paths because it alone was a repeat player in irrigation. Unlike the Old Regime, which lacked the power to intervene in local affairs, nineteenth-century governments had the incentives and ability to monitor water sales. Projectors could benefit from state-enforced prices by creating an *association*, which would announce rising prices for irrigation rights. In turn, rising prices would induce individual landowners to buy irrigation rights before the project occurred. Moreover, given the *association*'s commitment to such prices, creditors were then willing to finance projects because revenue was no longer risky.

These *associations* were also useful for solving the holdout problem in drainage negotiations. In marsh drainage, individuals held out so as to increase their share of the drained marsh. In the absence of state enforcement, it was not credible for projectors to announce that they would not change the allocation scheme for the drained marsh if someone held out for a larger offer. *Associations,* unlike projectors, were committed to simple equal division principles because those were the only ones the state would allow.

Again, the solution to the commitment problem was attained, not privately, but publicly through a complete reform of the process. This solution had to await the destruction of local *privileges* because local organizations would have seen the direct oversight of a canal company as an interference with their autonomy.[35] Thus, once again it appears that the Revolution was necessary to achieve this reduction in the transaction costs of providing irrigation.

———

In every case that we have examined, the Revolution reduced the uncertainty in property rights by altering the incentives of those individuals who controlled the process of defining or exchanging property rights. Problems associated with the lack of credible commitment were acute in Old Regime France. They persisted, most often, because internal solutions were unavailable. Solutions to the problem of durability were achieved by a simultaneous strengthening of the central government's power to act and to tax and an explicit division of authority among branches of government. Centralization gave the government oversight powers on irrigation and drainage projects. Meanwhile, the separation between the judicial and the financial branches of government reduced the incentives of the judiciary to prolong litigation.

To be sure, the Revolution resulted in wholesale redistribution of property rights. For example, lords lost all claims to communally owned land. In addition, some canals were confiscated by the state. Yet after the Revolution, property rights were more secure because the individuals and organizations who defined these rights had no incentive to alter them. Strengthened property rights had both direct and indirect effects on irrigation and drainage. The direct effects were associated with state oversight, which dramatically reduced the transaction costs associated with allocating the returns to development. The indirect effects involved the nearly total absence of litigation over property rights after the Revolution. The intimate relationship between the state's need for funds and the persistence of inefficient institutional arrangements give us yet another perspective with which to interpret the Revolution. As noted by David Bien, in times of crisis venal officeholders were highly flexible sources of credit for the state.[36] The state could not dispense with venality if it did not reform the tax regime. Thus, the tax crisis came to play a central role in causing the Revolution.

Tax reform, when it was begun, threw open the question of who should control the government budget, the Estates in a Parliament-like fashion or the king. As John Brewer as well as Douglass North and Barry Weingast have argued, the tax problem in England was solved through a transfer of power from the king to Parliament – a glorious revolution in-

---

[35] The most jealous guardians of local autonomy were, of course, judicial officers.
[36] Bien (1987).

deed.[37] In France, the problem of government revenue was solved in a tumultuous revolution. Whether or not the Revolution was necessary to solve the tax problem is an open question. Yet it seems beyond question that the reforms that reduced transaction costs in irrigation and drainage could not have occurred in the absence of a solution to the tax question.

[37] Brewer (1988, chap. 1); North and Weingast (1989).

9

Settlement, litigation, and the drainage of marshes in England and France, 1600–1840

Historians and economic historians have long wondered what accounts for the systematic differences between French and English agriculture. These differences encompass not only productivity, where it has long been recognized that British agriculture was superior to that of other preindustrial European countries, but also institutions. While recognizing that the determinants of agrarian achievement are complex, this chapter uses a theoretical model to argue that different legal institutions explain the divergence in the agrarian histories of France and England. As we have seen, institutions played a significant role in blocking investment in water control in Old Regime France. Here we explore how British institutions promoted one of the most dramatic features of agrarian development – the conversion of wasteland to farmland. One important way to reclaim waste and increase usable acreage was to drain marshes, a process that in England began before the seventeenth century and continued past the eighteenth. In France, by contrast, little drainage occurred before the Revolution.

As we have already seen, although marshes were most often called wastes, they played an important role in the rural economy. In low-lying areas, marshes were used by villagers for pasturing their animals and gathering forage and reeds, as well as for fishing. Villagers used the marshes as part of their commons, but they did not necessarily own the marshes. In fact, both the village as a community and the lord of the village had rights to the marsh. Even if the lord was most often the ultimate owner of the marsh, he could not always make unilateral decisions about its drainage. Indeed, authority over drainage was a complex legal issue. The importance of legal rules on marsh reclamation is not transparent from the archival records because litigants take those rules as given and concern themselves with the gains or losses associated with legal conflict. To the economic historian interested in a comparison between Britain and France, however, the consequences of variations in legal rules may prove paramount. Game theory allows us to go beyond the perspective of in-

149

dividual litigants and analyze how different burden-of-proof rules affected the development of agriculture. Focusing on a specific set of historical events strengthens the game-theoretic analysis because it demands that the settlement and litigation process be modeled in a richer and more detailed fashion than has heretofore appeared in the literature.[1]

Agricultural historians on both sides of the English Channel agree that the extension of cultivated areas through drainage and clearing was the most important source of output gains in early modern agriculture. Joan Thirsk argues:

The increase in [England's] agricultural output in the two centuries before 1650, however, probably owed less to improvements in productivity than to extension in the cultivated area. . . . Later as the area of easily accessible land diminished, attention was increasingly turned in marshy regions to reclamation by drainage.[2]

This pattern of marsh drainage in Britain continued into the seventeenth and eighteenth centuries. Interestingly, it was not associated with significant amounts of litigation. In France, by contrast, little drainage seems to have occurred past the thirteenth century, and until the 1750s one finds few attempts to drain marshes.[3] Thus, part of the task of this chapter will be to explain the sudden change in the attractiveness of wetland reclamation to French entrepreneurs. Despite the dramatic increase in the number of individuals attempting to drain marshes in the closing decades of the Old Regime, none seem to have succeeded. Most of the marshes that remained unimproved in France throughout the seventeenth and eighteenth centuries were eventually drained in the first half of the nineteenth century.

The historical pattern of drainage in England and no drainage in France seems odd. Indeed, we do not know of any significant difference in relative prices or factor endowment that should have led to greater development in England than in France.[4] Moreover, our investigation of drainage in Normandy indicates that some nineteenth-century French projects would have been quite profitable if undertaken in the eighteenth century. In England, the relative absence of litigation associated with reclamation projects runs counter to the intuition that, in the face of uncertain property rights, the judiciary will play an important role in allocating the benefits and costs of projects. In France, by contrast, much litigation took place, and yet no drainage occurred. This seems inconsistent with the

---

[1] Burden-of-proof rules, for example, have tended not to be analyzed. See, however, Sobel (1985) and Rubinfeld and Sappington (1987).

[2] Thirsk (1967, Vol. 4, 607).   [3] See Chapters 4 and 6.

[4] In fact, the scarcity of pasture in France should have made drainage more profitable there than in England. See, e.g., the reports of provincial officials in AN H[1] 1488–90.

view that litigation would lead to well-defined property rights and hence to drainage. In order to explain the odd pattern of litigation and drainage in England and France, it is necessary to take a closer look at the role of marshes in the village economy and at the distribution of property rights over marshes.

For ease of exposition, let us consider the problem within a village, which for our purposes is inhabited by peasants who form the community; assume further that this village has a marsh and that the lord of the village wishes to drain that marsh. In most regions of England, as in France, the lord originally held title to all the land in the village, and a variety of arrangements allowed him to gain revenue from that land. These ranged from the outright sale or rental of land to direct production through the various levels of serfdom. Each one of these arrangements was in effect a grant of farmland from lord to peasant. Associated with each grant of farmland were use rights to the land that remained un-farmed.[5] As a whole, the unfarmed land in the village was known as the common. In fact, the unfarmed land must be divided into two categories: first, the village commons (hereafter, the commons), a certain amount of pastureland that the lord was obligated to provide to the village; second, the waste, which was the residual unfarmed land in the village. Although there is no doubt that the legal owner of the commons and the waste was the lord, his authority over the commons was quite small. In fact, any change in the use of the commons required the approval of the village community. The authority of the lord over the waste was complete, provided that he had not transferred property to a third party or to the village.

Until drainage occurred, there was little to distinguish a marsh that was part of the waste from the commons; both were equally accessible to villagers. After drainage, the value of the marsh would be greatly increased, and the lord would want to separate it from the commons – in effect, enclose the drained land and rent it out. The crucial legal issue concerned when the lord had the right to drain the marsh over the objections of the village or, in other words, under what circumstances the village could block the enclosure of the waste. If drainage had occurred regardless of whether the marsh was determined to be waste or common, the issue of attributing the property rights to the marsh would be of little economic interest. The distribution of property rights, however, was crucial to the success of drainage. On the one hand, when the lord had the authority to decide upon drainage alone, and when he collected most of the surplus, we would expect reclamation to occur. On the other hand, when the marsh was declared common, drainage was much more difficult because its success depended on the agreement of some or all of the

---

[5] Ault (1972, 16).

villagers. One might assume that villagers and lord would have reached an accord to share the benefits of reclamation, but historically, as we saw earlier, this was not the case.

One frequent barrier to drainage was that villagers held out for larger shares of the surplus generated by improvement. Theoretical support for intravillage bargaining failures can be found in a paper by Mailath and Postlewaite.[6] In that paper, a surplus is to be produced and divided among the members of a group. Each individual has a privately known reservation value below which he or she does not want to participate in the production and division of the surplus – in our case the drainage of the marsh. Each individual must announce his or her reservation value knowing that the surplus to be created is likely to be a good deal larger than the total of the expected reservation values. Mailath and Postlewaite show that all individuals have an incentive to free-ride and inflate their reservation values. The result is that, as the number of participants grows, each individual increases his or her announcement until it reaches the highest possible reservation value. The sum of these demands is greater than the surplus and thus the project fails. Consistent with their analysis and the historical evidence discussed in Chapters 4 and 6, we will assume that transaction costs made it very unlikely that marshes would be drained if they were declared common land.[7]

Had property rights been well defined, the judiciary would have played no role in this chapter of the agrarian history of England and France. However, property rights to marshes were uncertain because the title to them had become obscured by the passage of time. As already noted, waste and commons had originally been indistinct in terms of access and use. Yet the village had access only to the common as a right, and frequently had to pay rent to the lord for the waste – a rent that varied with the value of the marsh. During the fourteenth and fifteenth centuries, some lords who had become strapped for cash sold control of their waste to villages.[8] Other lords had let the administration of their estates lapse, and the villagers had been able to enjoy the use of the waste free of charge. In short, some lords surrendered the waste to the village while some villages usurped the rights of the lords. As a result, by the seven-

---

[6] George Mailath and Andrew Postlewaite, "Asymmetric Information Bargaining Problems with Many Agents" (Univ. of Pennsylvania, mimeo, 1988).

[7] Intravillage bargaining successes seem to have been crucially dependent on the social choice rule that prevailed in the village. In England medieval statutes transformed the unanimity rule into a majority rule, while in France unanimity was required until the 1820s. Not surprisingly, more common marshes were drained in England than in France. For French evidence on intravillage disputes, see AN, H^1 1492–1500, and AD Calvados, C 4201.

[8] Either the marsh was sold outright or perpetual access was granted in return for a lump sum payment and a small annual fixed fee.

teenth century it was rarely possible to distinguish the waste from the commons.

Yet if drainage was to proceed, it was necessary to determine who had title to the marsh. In favor of villagers was the fact that the marsh had come to be treated as part and parcel of the commons, giving villages whatever rights occupancy conferred. In favor of the lord was his original ownership of the land. To decide the issue, both sides debated the validity of medieval contracts. The problem with these contracts was that some of them had been lost, some could be fakes, and some could relate to another part of the common land. A village opposing a drainage scheme would bring the issue to court, and thus the question of property rights was to be resolved judicially.

———

The judicial confrontation between lord and village was fraught with uncertainty because, as I will argue, the evidence available to either party was scant and rarely conclusive. In order to model the effect of uncertainty on the judicial process we must first go deeper into the issue of evidence; second, we will present the stylized version of the judicial environment.

Because a very long time had passed since the beginning of contractual exchanges between lord and village, it was unlikely that they would have retained copies of the same set of contracts. In fact, one can argue that village and lords had asymmetric information. The ideal pieces of evidence for lords were long series of rental payments by the village that varied with the market value of the marsh. A varying level of rent showed that the lord retained authority over the marsh and thus probably ownership as well. Because the ownership claims of the lord were most often based on rental payments, it seems likely that the village would know what evidence was available to the lord if the case were to go to court.

The ideal evidence for a village was a contract that explicitly transferred the marsh to the common. In the absence of such an explicit contract, other contracts in which the lord recognized the village's ownership of the marsh would be valuable evidence. Although both the lord and the village originally had copies of these contracts, the village was more likely than the lord to preserve such a parchment because they were more valuable to the village.[9] Moreover, it would be unlikely that the lord would know exactly what contracts had been preserved by the village. Furthermore, some villages probably preserved more documents than others.

---

[9] One should also note that the evidence presented in French cases of marsh drainage was precisely contracts of sale or transfer for villages and the rental receipts for lords; AD Calvados C 4200–5.

Those with more careful archivists would no doubt have had a stronger hand in court.

The facts outlined above about evidence suggest that there was asymmetric information, that villages knew what evidence was available to the lord, and that the lord did not fully know what evidence was available to villages. Given the structure of information, it is clear that villages had little to gain from procedural rules such as discovery. For lords, however, discovery could be an important means of extracting information from the village. I model discovery as a one-time choice for the lords to ask the community a fixed set of questions about some contracts potentially held by the village. The lord may not discover all these contracts, so discovery will succeed in reconciling the information only part of the time. If discovery succeeds, settlement occurs with probability 1 in this simple model because village and lord share the same beliefs about the court outcome, and both prefer to save on court costs. If discovery fails to reconcile the positions of lord and village, the asymmetry in information remains.

In case of trial, neither party to the suit can accurately predict the court's action. Both parties know what contracts they have access to and may also discover the evidence held by the opposition. The litigants, however, do not know with certainty which contracts the court will find valid and which it will reject. In short, what matters for lords is to convince the court that the marsh is waste and for the villagers to demonstrate to the court that it is common. For simplicity, I assume that the marsh is either common, in which case it is impossible for the lord to convince the court that it is waste, or that the marsh is waste, in which case it is impossible for the village to convince the court that it is common.[10] The ability of village or lord to convince the court depends on legal expenditures (i.e., resources spent on presentation, cross examination, and legal expertise). Higher expenditures by the village increase the likelihood that, when the village is in the right (the marsh is common), it succeeds in convincing the court that the marsh is common; the same applies to the lords' expenditure when the marsh is truly waste. Even if a lord or a village is in the right, however, it is not guaranteed that the court will be convinced, and an unconvinced court applies a burden-of-proof rule.

The final element of the model is the ability to settle out of court. Clearly, if there was much drainage in England but no litigation, then out-of-court determination of property rights must have occurred frequently. For simplicity, I collapse the out-of-court bargaining process into a single out-of-court settlement offer made by the lord. The only reason

---

[10] Since all marshes had originally been waste and marshes became common on the basis of contract, such a simple division of cases seems appropriate.

to allow the lord to make the final settlement offer is that lords usually initiated formal drainage procedures by presenting villages with a drainage contract that described the division of property rights after the completion of drainage – in effect, a settlement offer.[11]

The model presented here builds not only on the stylized facts just presented, but also on the game-theoretic treatment of settlement and litigation developed in the literature of law and economics.[12] The two articles most relevant to my analysis have been written by Lucian Bebchuk and by Jennifer Reinganum and Louis Wilde.[13] These authors focused on the impact of different rules for allocating litigation costs between plaintiff and defendant on equilibrium trial possibilities. Thus, their articles did not consider that judicial outcomes are typically unpredictable in large part because documenting claims is uncertain and expensive. To understand more fully the effect of litigation costs on the judicial process, one has to examine the relationship between legal expenditures and outcomes, burden of proof and settlement, as well as discovery and the disposition of cases.[14]

For clarity in the next few sections, which are devoted to the formal model, let us ignore the history of drainage in England and France. The historical implications of the model will be discussed at the end of the chapter. In keeping with the details of marsh drainage, the model involves two parties, a lord (defendant) and a village (plaintiff). On village land, there is a marsh that provides pasture to the villagers at no cost. The value of that pasture is P. The lord wants to drain the marsh, but it may belong to the village commons. If the marsh is found to be

[11] The results of the analysis in terms of drainage would not differ substantially if the villages were to make the final settlement offer. However, it is a different game-theoretic problem entirely; cf. Reinganum and Wilde (1986).

[12] Those readers not inclined to follow the technical details may wish to skip the next section. However, I believe that the theoretical analysis and the historical discussion that follows should be taken as a whole.

[13] Bebchuk (1984); Reinganum and Wilde (1986). See also Meurer (1989). The key difference between their models is that Bebchuk has the uninformed party make the settlement offer, while Reinganum and Wilde give that opportunity to the informed party. Whether the informed or the uninformed party makes the offer affects the probability of settlement as well as the magnitude of the expected transfer between plaintiff and defendant. While Bebchuk's model is closest to the one developed herein, it is too sparse in institutional details to afford insight on the role of burden of proof in accounting for the different histories of French and English marshes.

[14] The model confronts these institutional issues in a specific historical context; yet the conclusions extend beyond the problem of allocating property rights between lord and village.

common, the village is entitled to decide the issue of drainage and the marsh remains undrained. The lord receives nothing and the village maintains its pasture. Profits in the absence of drainage are zero for the lord and the village. If the marsh is found to belong to the lord, it will be drained and its value will then be π; we assume $\pi > P$.[15] The village will receive a small compensation (α) less than the value of the free pasture, so this outcome will lead to a loss for the village of $\alpha - P$ and a profit for the lord of $\pi - \alpha$.[16] Villages can be of two types: strong (denoted by a subscript s) or weak (denoted by a subscript w); the probability that a village is weak is denoted p. In any village the probability that the marsh is common is β_i, and the probability that it is waste $(1 - \beta_i)$. The probability that the marsh is common is greater in strong villages than in weak ones $(\beta_s > \beta_w)$.

Assume that village strength is determined by a random process and that only the village knows whether it is strong or weak.[17] The village is only partially informed about its right to compensation because its classification by type is only the probability that the marsh is really common, not what the court will decide. The lord knows neither the true allocation of the marsh (waste or common) nor the village's type, but he knows the probability distribution of types of villages and the probability that the marsh is common given the different types. This structure makes the village better informed than the lord, in keeping with the earlier discussion, while maintaining that both parties are uninformed about the true nature of the marsh, in keeping with the irregular record keeping of the early modern period.

In the next stage of the process, the uninformed lord applies a discovery rule that reveals the type of the village with probability δ. Discovery is modeled as a process of asking questions before the settlement offer is made. There is no need to model the issue of discovery costs, since they are all borne before the out-of-court settlement offer is made; as a result, all discovery costs are sunk.[18] With probability δ, the lord discovers the set of contracts owned by the villages that bear on the issue of marsh ownership. In this case, the lord knows the type of the village and makes

[15] For simplicity, the model neglects the fact that lords as well as villages enjoyed pasture rights.
[16] The assumption that π is greater than P is only meant to suggest that ex ante marsh drainage was thought to be profitable. In fact, in many cases it was unsuccessful. See Summers (1976) and Darby (1983) for evidence that on the eastern edge of England many drainage schemes failed.
[17] Technically, in the first stage of the process the village's type is decided by a random draw, which is revealed to the village.
[18] Discovery cost would be borne primarily by villages, and they would benefit the lord. Thus, the lord would always demand that the village expand the maximum resources possible in discovery. Because we focus on sequential equilibria, past expenditures do not directly affect future decisions.

Table 9.1. *List of variables*

P	Value of the undrained marsh to the village
π	Value of the drained marsh to the lord
α	Compensation that villages receive when lords repossess wasteland
β_i	Probability that the marsh is common for village of type i (s for strong, w for weak)
p	Probability that the village is weak
q	Lord's beliefs about the village's type
S	Settlement offer from the lord to the village
k_i	Probability that a village of type i sues in court
F_v, c_i	Fixed and variable court costs for villages
F_l, c_l	Fixed and variable court costs for lords
t	Burden-of-proof rule
$E_v(\cdot)$	Probability of finding convincing evidence for villages
$E_l(\cdot)$	Probability of finding convincing evidence for lords
W_i	Probability that the lord wins in court
$d(t)$	Probability that the marsh is drained given a burden-of-proof rule of t

a settlement offer (denoted S) equal to the expected value of trial for a village of that type. The offer is accepted with probability 1. With probability $(1 - \delta)$, discovery fails and the lord must make his settlement offer uninformed.

If the lord must make an uninformed settlement offer, the village decides either to sue in court (with probability k_i) or to accept the offer (with probability $1 - k_i$). If the village chooses to accept the offer, the game ends; if it sues, a trial occurs and both parties bear court costs. Going to court involves both fixed and endogenous costs. The fixed costs of the village are denoted F_v, those of the lord F_l. Fixed costs are intended to capture such things as the cost of delay and the fixed legal expenses associated with the trial. The variable costs for villages of type i are denoted c_i, those of the lord c_l. These are the costs of preparing arguments, presenting the evidence, and cross examination. Because the model is designed to emphasize the distinction between the true nature of the marsh (waste or common) and the ability to demonstrate the truth, I have introduced the issue of evidence and the necessity to convince the court. All the parameters in the model are described in Table 9.1.

In the case of litigation, both parties are able to hire lawyers to do research for them and to present evidence to the court. The court makes a decision about the village's right to the marsh based on the evidence brought to the trial and the strength of the arguments presented by the lawyers. Based on their parties' evidence and on cross examination, either the lawyers will convince the court or their efforts will be inconclusive. In this model, judges and courts do not try to infer the original parameters of the case (p, β_w, and β_s); rather, they try to infer the true state of the marsh (common or waste) from the arguments presented by the lawyers. Thus, courts care only about the arguments presented by both parties. For simplicity again, I assume that a litigant's lawyer can act only in his favor. So if a village has evidence that the marsh is waste, the village's lawyer will not report that evidence. The lord's lawyer, however, may through examination bring that evidence to light. I also do not allow for the possibility that the litigants can mislead the court. Thus, in the case of common marshes the court can only be convinced that it is a common marsh. These assumptions allow me to simplify the model so that the efforts of a party's lawyer matter only when that party is in the right. I can therefore separate the effects of village (plaintiff) expenditures and the effects of those of the lord (defendant).

If the court is convinced by either party that the marsh is common, then the court awards the property rights to the village (lord); if, however, the court is convinced that the marsh is waste, the court awards the property right to the lord. If, after the closing arguments, the court remains unconvinced, the court awards the marsh to the village with probability t. Variations in t are equivalent to variations in the burden-of-proof rules. Indeed, burden-of-proof rules come into play only when the court remains unconvinced. For example, if t is 0, the burden of proof rests completely on the village. Since there is no change in revenue when the court finds that the marsh is common, applying the burden-of-proof rule leads, in expected value, to a loss for the village of $(1-t)\alpha - P$ and a profit for the lord $(1-t)\pi$.[19]

For cases in which the marsh is actually waste, the probability that the lord convinces the court that it is actually waste, given expenditure c, is defined as a concave and differentiable function $E_v(c)$. The probability of convincing the court that it is common is 0. For cases in which the marsh is common, the probability that the village convinces the court that it is common, given expenditure c, is defined as a concave and differentiable function $E_l(c)$. The probability of showing that the marsh is waste is 0. Furthermore, the conditions sufficient to avoid the uninteresting equilibrium in which neither player expends any legal resources are

$$\frac{\partial E_v(0)}{\partial c} > \frac{2}{\beta_s(\alpha - P)} \quad \text{and} \quad \frac{\partial E_l(0)}{\partial c} > \frac{2}{(1-\beta_s)\pi}.$$

[19] This specification of burden of proof neglects the fact that case strength is a continuous rather than a discrete variable. In this alternative perspective, burden-of-proof rules define cutoff points for the allocation of cases between plaintiff and defendants. While this more plausible specification may appear more plausible, within this model it leads to exactly the same results as the one chosen here. Since the model is already complicated, it seems unnecessary to further burden ourselves.

Settlement, litigation, and marsh drainage

Because convincing the court is a probabilistic event, even when the marsh is truly common it is possible for the court to remain unconvinced. If the burden-of-proof rule is not fully on the lord, the court may decide the marsh is waste (an event that occurs with probability $(1-t)$). The problem is now sufficiently well specified that we can write down the expected returns to trial for the village and the lord explicitly. Given a plaintiff type, the probability that the lord wins and the marsh is declared waste is

$$W_i(c_i, c_l) = \beta_i(1 - E_v(c_i))(1 - t) + (1 - \beta_i)(E_l(c_l) + (1 - E_l(c_l))(1 - t)).$$

The expected value of going to court for the village is

$$\Lambda_i(c_i, c_l) = P - W_i(c_i, c_l) \ (P - \alpha) - c_i - F_v. \tag{1}$$

The lord's revenues given the village's type are

$$\Psi_i(c_i, c_l) = W_i(c_i, c_l) \ (\pi - \alpha) - c_l - F_l. \tag{2}$$

Equations (1) and (2) specify the allocations of marshes between waste and common in the absence of settlement opportunities. As the latter part of the chapter will make clear, in seventeenth-century France the state discouraged out-of-court settlements of property rights questions. Marshes were therefore allocated by judicial decision. Proposition 1 suggests that burden-of-proof rules play a crucial role in the disposition of cases when there are no settlement opportunities.

Proposition 1. In the absence of settlement opportunities, the burden-of-proof rule will allocate strictly more marshes to the commons as t increases; thus, if t is 1, it is least likely that marshes will be drained.

In England as well as in eighteenth-century France, settlement opportunities were available. To examine these cases, we must specify the objectives of the lord and the village if out-of-court settlement is possible. The expected value to the lord of going to court is

$$\upsilon_l(c_l) = q\Psi_w(c_l) + (1 - q)\Psi_s(c_l). \tag{3}$$

Here, q is the lord's belief about the probability that the village is weak if the village has refused the settlement offer. If out-of-court settlement is possible, then the expected value of deciding whether the marsh is common or waste for a village of type i is

$$\begin{aligned}V_i(k_i, c_i) &= (1 - k_i)(S - P) + k_i\Lambda_i(c_i)\\ &= (1 - k_i)(S - P) + k_i(P - W_i(c_i, c_l)(P - \alpha) - c_i - F_v),\end{aligned} \tag{4}$$

while the returns to deciding whether the marsh is common or waste for the lord are

Property rights and litigation under absolutism

$$\Pi_l(S, c_l) = (p(1-k_w) + (1-p)(1-k_s))(\pi - S) + pk_w\Psi_w(c_l) + (1-p)k_s\Psi_s(c_l)$$
$$= (p(1-k_w) + (1-p)(1-k_s))(\pi - S) + pk_w(W_w(c_w, c_l)(\pi - \alpha)$$
$$- c_l - F_l) + (1-p)k_s(W_s(c_s, c_l)(\pi - \alpha) - c_l - F_l). \tag{5}$$

To complete the formal structure of the model, we must define the strategies of the village and the lord and choose an equilibrium concept.

Definition 1. A strategy for the lord is a settlement offer S, and a level of legal expenditures c_l that may depend on S. A strategy for a village of type i is a probability that the village refuses the settlement offer k_i that will depend on S, and a level of legal expenditures c_i that may depend on S and k_i. Although the village may pursue a mixed strategy in deciding whether to accept the lord's offer, the lord only observes the outcome of the village's action. In equilibrium, of course, the village's actions will have to be incentive compatible – that is, the village will randomize between litigating and accepting the settlement offer only if it is indifferent between these two courses of action. All equilibrium values will be superscripted with an asterisk.

Definition 2. A sequential equilibrium of this game has four elements: (1) a set of legal expenditures c_l^*, c_w^*, c_s^* that maximize the returns from a trial ($\Lambda_i(c_i)$ and $\Lambda(c_l)$); (2) a pair of litigation decisions k_w^*, k_s^* that are optimal given an offer of S^* and the expected returns from the trial; (3) a settlement offer S^* that minimizes the lord's cost of sorting the property rights given what each party expects to get at trial ($\Lambda_i(c_i^*)$ and $\Lambda(c_l^*)$); (4) a lord's belief about the villages type, q^*, that satisfies Bayes's rule: $q^* = pk_w^*/(pk_w^* + (1-p)k_s^*)$. In short, the equilibrium requirement is that at each stage of the process equilibrium strategies must be both optimal and incentive compatible.

Theorem 1. There exist at most two equilibria in this game, E_1 and E_2. In E_1 weak villages randomize between going to court and settling, and strong villages all go to court. In E_2 all villages settle out of court. The equilibria are characterized by the following:

(1) Optimal levels of expenditures for villages are given by

$$\frac{\partial E_v(c_i^*)}{\partial c} = \frac{1}{\beta_i(1-t)(P-\alpha)}, \qquad i = s, w;$$

(2) Optimal expenditures for the lord are given by

$$\frac{\partial E_l(c_l^*)}{\partial c} > \frac{1}{[q^*(1-\beta_w) + (1-q^*)(1-\beta_s)](t)(\pi - \alpha)}.$$

(3) Optimal litigation probabilities for each equilibrium are as follows: $k_w^* \in [0, 1]$ and $k_s^* = 1$ if $S = S_1^*$, $k_w^* = k_s^* = 0$ if $S = S_2^*$.

160

(4) Optimal settlement levels S_1^*, S_2^* (corresponding to E_1, E_2) are defined by the following: S_1^* maximizes the lord's revenues when strong villages sue with probability 1, and $S_2^* = \Lambda_s(c_s^*)$.

(5) The Bayesian beliefs of the lord are $q^* = pk_w^*/(pk_w^* + (1-p))$ if the equilibrium is E_1, and $q^* = 0$ if the equilibrium is E_2. All proofs are given in Appendix 3.

The game features two types of equilibria because lords and villages have different information. Since trials are costly, the lord always wants to settle with weak villages. The lord would also like to settle with strong villages, but since he cannot distinguish them from weak villages, he faces two options. The lord may compensate all villages at a high settlement level. He will do just that when his litigation costs are very high or when the probability that villages are weak is very small. Under such conditions, the lord will maximize his profits by making a high settlement offer, which leads to a type E_2 equilibrium.

Alternatively, when litigation costs are low, the lord may make only a low settlement offer. In a type E_2 equilibrium, some litigation takes place because strong villages sue with probability 1, while weak villages randomize between accepting the lord's offer and going to court. In this case, the lord may not become informed about the villages' type, because weak villages may randomize.

Corollary 1. All equilibria are locally unique; that is, for any set of initial conditions such that both E_1 and E_2 are equilibria, there exists a set of initial conditions that are arbitrarily close such that only E_1 or E_2 is an equilibrium.

Corollary 1 allows us to derive comparative statics results for both equilibria. Since the comparative statics results are most interesting in the case of E_1, we shall focus on it. All the results (except for those concerning evidence expenditures and litigation probabilities) also apply to the other equilibrium, and because they can be derived straightforwardly, no proofs are given. As with most of the literature, in this case higher village fixed court costs lead to lower settlement offers. Indeed, when weak villages face higher fixed court costs, they will be willing to accept a lower settlement offer. Increasing village fixed costs also leads to a higher acceptance probability. Higher lord court costs increase the magnitude of the settlement offer and also raise the probability that the offer will be accepted. Thus, as court costs rise, litigation falls.

An increase in the value of pasture on the undrained marsh (P) leads to a higher settlement offer, a higher probability of litigation, and greater expenditure on evidence by the village. Increasing the value of the drained marsh (π), while holding the value of the undrained pasture constant, results in an increase in the litigation expenditures of the lord and may reduce the probability of litigation or the size of the settlement offer. Indeed, if π increases, the lord is willing to invest more in order to win a suit. Since the lord is willing to invest more in litigation, this makes going to court less attractive for weak villages.

Definition 3. Let $V_i(\tau)$ be the expected value of going to court for a village of type i when the burden-of-proof rule is τ, and let $c_j(\tau)$ be the optimal level of expenditures for litigant j given a burden-of-proof rule of τ.

Proposition 2. If $\partial c_1(\tau)/\partial t < p(\partial V_s(\tau)/\partial t - \partial V_w(\tau)/\partial t)$ for all τ in $[0-1]$, then there exists a unique t^* such that if $t < t^*$, then the equilibrium selected is E_2, and if $t > t^*$, E_1 is selected.

In words, Proposition 2 will hold if a fall in t reduces the lord's court costs to a lesser extent than the cost of not differentiating between weak and strong villages. This condition is likely to hold in the case of marsh drainage for two reasons. First, judicial procedure guaranteed that a small proportion of judicial costs would be controlled by the litigants. Thus, variation in t would have little effect on judicial expenditures. Second, the high degree of uncertainty in the validity of contracts suggests that courts would most often remain unconvinced by either party's evidence independent of the level of legal expenditures by lord and village. If the court were unlikely to be convinced by the evidence, the overall investment in legal expenses would be small relative to fixed court costs or to the difference between possible awards. Changes in the burden-of-proof rule would nonetheless have a large impact on the difference between the value of going to court for strong and weak villages, because cases were most often decided by burden of proof.

Beyond Proposition 2, a few other comparative statics results illuminate the importance of burden of proof. As the burden-of-proof rule becomes more favorable to the lord (equivalently, as t is decreased), the lord will lower his settlement offer — because the probability that the marsh will be declared common also falls. Thus, making the burden of proof bear more strongly on the village strengthens the hand of the lord not only in court, but also in the settlement stage. Finally, as t falls, the evidence expenditures of villages of either type rise, while the expenditure of the lord falls. Indeed, in this model, as the court's rule for allocating undocumented cases becomes more favorable to the lord, his incentive to expend resources on lawyers falls.

The impact of discovery on settlement is clear: As the probability of discovery rises, the ex ante expected settlement to weak villages falls and the expected settlement to strong villages rises.[20] In the case of a weak village, the ex ante settlement offer falls as discovery rises. If discovery succeeds, the lord knows the village is weak, so he is willing to devote more resources to arguing in court than if there is a positive probability that the village is strong. Higher legal expenditures by the lord reduce the expected court award for the village and thus reduce the settlement offer. Therefore, a weak village is always offered less when it has been identified than when discovery has failed. So an increase in the probability of discovery helps the lord to discriminate between weak and strong villages, lowering the expected compensation to weak villages and increasing the expected compensation to strong ones.[21]

The histories of drainage in England and France provide good opportunities to test this model. One of the model's conclusions is that burden of proof has an important impact on the settlement level and the probability of out-of-court settlement. First, because the burden of proof on the question of drainage varied between France and England, the model yields different predictions for each country. Second, in the case of France, the model predicts a change in the pattern of activity – litigation and drainage – in the last half of the eighteenth century after the enactment of reforms designed to ease the constraints on out-of-court settlements.

The stylized facts presented earlier on the history of drainage in England suggested that drainage occurred in the absence of litigation. Historians have also noted that there were regular, if not frequent, occurrences of violence surrounding the drainage of marshes from the commons.[22] The absence of litigation suggests that most property rights were allocated out of court. In effect, England corresponds to an E_2 equilibrium in the model – one in which all villages are compensated outright. Such an equilibrium is likely to arise if the difference in the expected value of trials between strong and weak villages is small enough relative to court costs that lords settle with everyone. The features that could have lowered the value of trials for English villages included a low value of marsh pasture (P) and institutions that reduced villages' probability of winning. As I argued earlier, since the agrarian technologies in England and France would have made the value of marsh pasture similar

---

[20] In an equilibrium of type 1 that offer is rejected with probability 1 by strong villages. However, if discovery succeeds, the lord knows the village is strong and so he will offer such a village a higher compensation.

[21] Changing the costs of discovery – which are borne by the party with information – does not affect either the compensation offer or the probability of trial because discovery costs are fully sunk at the settlement stage.

[22] Thirsk (1967, Vol. 4, 223); also Summers (1976, 104–9), Darby (1983, 57–61, 67), and Lindley (1982).

in both countries, we must look to institutions to understand what made trials unappealing to English villages. In fact, trials were unattractive to English villages because the villages bore the burden of proof in cases dealing with ownership of marshland. The historical evidence for this is compelling. In the statute of Merton (1235) and again in the statute of Westminster (1285), Parliament and the king reaffirmed that villages were entitled to limited pasture only and that wastes were the private property of lords.[23] These statutes remained the basis of the law regulating common and waste until well into the nineteenth century. Moreover, in the absence of a contract explicitly transferring marshes from waste to common, the village was not to interfere with any reclamation project.

Putting the burden of proof on the village corresponds to $t = 0$ in the model. In this case, not only was the village sure of losing all the cases in which the marsh was actually waste (with probability $1 - \beta_i$), but the village would lose all the cases in which the marsh was common unless it could convince the court that it had a valid contract to the marsh (a probability equal to $\beta_i(1 - E_i(c_i))$). Given the requirement for contractual evidence, the fact that villagers used the marsh for pasture was of little help to the village. In this case, it was very unlikely that the village could convince the court. Formally, when t is zero, the value of going to court is minimized for all villages $(\Lambda_i(c_i) = (\alpha - P) + P(1 - \beta_i E_i(c_i)) - c_i - F_v)$, a fact that reduces the expected transfer from lord to village independent of which equilibrium is selected. More important, Proposition 2 suggests that, when the village bears the burden of proof, an E_2 equilibrium is the most likely outcome (out-of-court settlement occurs with probability 1). Thus, it was possible for the lord to compensate a village at far less than the value of the pasture, by, for example, transferring a small portion of the waste to the common.

By the seventeenth century, the medieval statutes that regulated the drainage of marshes had been tried and tested, and villages knew quite well the limited extent of their property rights. A measure of the importance of contract law over the commons appears in Keith Lindley's book on the Great Fen:

Sir John [Mowbray] as the then manoral lord [of Epsworth], in return for their consent to his enclosure of part of the commons granted the commoners the remainder free from any further improvements by the lord or his successors. This agreement was enshrined in an indenture dated May 1359, a document treasured by the commoners and carefully preserved in the parish church of Haxey in a specially prepared chest bound with iron. . . . The whole parish was kept regularly aware of the document's existence as the chest was placed under a window which depicted Sir John Mowbray holding a document, commonly reputed to be

---

[23] Rothwell (1974, 352, 455–6). For nineteenth-century interpretations, see Maidlow (1867) and Cooke (1864).

the indenture. . . . Hence [in 1629] Epsworth commoners were fully conscious of their unassailable legal position on the question of title.[24]

It seems clear that the villagers felt that the strength of their title rested directly on their ability to produce the physical contract that Sir John Mowbray had granted them. The very strength of the commoners of Epsworth underscores the weakness of most other villages, whose contracts, if they had ever existed, were not preserved with such care and were often lost. As a result, the villagers' legal means of protecting their access to the marsh had been closed off, and they resorted to extralegal means of pressure on the lord and the king to redistribute the surplus from drainage: They revolted. But because the king was one of the largest landlords in England, he had no interest in hearing the complaints of villagers and changing the law.[25] So the revolts were put down, and more important, the marshes were drained with little litigation.

In France, however, things were never so simple. First, the burden of proof in the case of marsh drainage was always against the lord.[26] Second, the king required that property rights be judicially determined, in effect ruling out the possibility of out-of-court settlements. As a result of the burden-of-proof rule, villages had a particularly strong hand. Unless the lord could convince the court that in the past thirty years the village had recognized by contract that the marsh was waste, it would be declared common. From Proposition 1 we know that this is the case in which drainage is least likely to proceed after litigation. Moreover, the French rules minimized the lord's profit from sorting out the property rights $(\psi(c_1) = (\pi - \alpha) \ (p(1-\beta_w) + (1-p)(1-\beta_s))E_1(c_1) - c_1 - F_1)$. One should note that, if fixed court costs are very high or if it is difficult to prove one's case ($E_1(c)$ is very flat), then $\psi(c_1)$ may be negative, and a lord would not want to go to court. This was precisely the case in seventeenth-century France because rent-seeking practices in the judiciary led to very high court costs, and the high degree of uncertainty about the validity of contracts also reduced the probability the lord would convince the court.[27]

French institutions seem to have made it very unlikely that the lord would win in court even if the village was weak, and thus drainage was unlikely to occur after the court decided the property rights issue. Had the village and the lord been able to settle out of court, we might presume that the profits from drainage would have been an incentive for lords and villages to reconcile their differences privately. Because the lord was in a

[24] Lindley (1982, 32).   [25] Thirsk (1967, Vol. 4, 260–76).
[26] See, e.g., AD, Calvados C 4197–4205, for cases in which villagers won a judgment on the basis that they had pastured their flocks on marshes without paying any explicit rent for the marsh.
[27] Derlange (1987, 53); Dewald (1987, 154, 262–3); AD Calvados, C 4226.

disadvantaged position if litigation occurred, the village would have been compensated at a high level, yet drainage would have occurred, and the burden-of-proof rule would have had only redistributive consequences. In France it was impossible to settle out of court; and worse, very high court costs made lords unwilling to invest in a judicial determination of property rights. Accordingly, in the seventeenth century no marshes were drained and there was no litigation.

Given the fiscal problems it faced, the Crown was reluctant to change the laws surrounding marshes.[28] Unlike the situation in England, in France lords were exempt from many land-based taxes. In northern France, this exemption was personal and was thus transferred to all land acquired by nobles. While the use of the marsh remained pastoral, the Crown was able to make the villagers bear some tax. However, if the land was transferred to the lord, such tax revenue would have been eliminated. This problem transcended the simple issue of marshes because some villages attempted to sell most of their land in order to reduce their tax burden.[29] To reduce the erosion of the tax base, the king insisted on a judicial determination of property rights over any potentially common land, and more specifically over marshes, and lay the burden of proof on the lord.[30] The likelihood of institutional change was further reduced by the fact that the king of France, unlike the king of England, did not own a significant portion of his country, and as a result had little to gain directly from successful marsh development.

As the eighteenth century progressed, concerns with total agricultural output led the Crown to attempt to revise the rules of marsh drainage. The Crown announced several times that it would support endeavors to drain marshes and that it would liberalize the out-of-court settlement process.[31] In effect, the Crown continued to require a settlement compensation in land. As noted in Chapter 4, the marsh could be drained only if it were waste, in which case the village would receive a small portion of the reclaimed land. In all other cases, the marsh was recognized as common, and drainage would be decided by the village after the rule for dividing the village's share among the inhabitants had been agreed upon.[32]

Whatever the royal government's intentions, the law accomplished little. The revised settlement rules led to systematic offers of one-third of marshes to villages, which in nearly all cases were refused by the villages.

[28] See Chapter 2, Mousnier (1971), and Esmonin (1913).
[29] Derlange (1987, 55–79); Hoffman (1986, 37–55).
[30] See Chapter 4 for more details.
[31] See the edicts of 1765 on division of the commons and those of 1766 that applied to marshes. Neither were ever enforced (AN H¹ 1486–1500, 1511–15).
[32] See AN H¹ 1496.

Settlement, litigation, and marsh drainage

The Crown's announced support for drainage projects can be translated into a shift in *t* away from 1, raising the probability that the lord would be awarded the property rights to the marsh. Both reforms should have raised the probability of success for drainage projects. In fact, a number of factors made the new law inoperative. Because villages could not agree upon internal division rules, a settlement in which the village received two-thirds of the marsh stated that the marsh would remain undrained. If only one-third of the marsh was offered by the lord to the village, however, only villages with weak property rights would accept such a settlement, and there were relatively few of those. Indeed, over the course of the sixteenth and seventeenth centuries, villages had been able to acquire squatter's rights to marshes by simply pasturing their animals on the waste for generations.[33] Thus, the settlement opportunities offered by the royal reforms were far too limited.

The announced change in the burden-of-proof rule could have increased the number of successful drainage projects by increasing the likelihood that the court would decide the marsh was waste. Yet these reforms depended on the ability of the Crown to control the judiciary. Despite the absolutist trappings of the Bourbon monarchy and the king's oft-proclaimed judicial supremacy, the Crown remained unable to impose any reform on the judiciary. Like many other reforms, the liberalization of drainage was viewed as an attack on *privileges*. Judges systematically opposed any change in the law that threatened *privileges*, and in drainage cases always sided with villages.[34] Lords and developers filed and litigated approximately thirty cases in Normandy between 1740 and 1789, but none led to drainage. There was only one instance of violence, when a small group of villagers destroyed the early phase of a drainage project. Because no one in the French administration was willing to investigate the issue, drainage was abandoned.[35]

Proposition 1 allows us to confront a final question: What would have been the pattern of development if, despite the impossibility of out-of-court settlements, France had had the same burden-of-proof rule as England? Because the property rights to marshes were very unclear, burden-of-proof rules played a crucial role in decisions about whether marshes were waste or common. If *t* was 1 as in France, marshes would be deemed waste with probability

---

[33] See, e.g., AN H¹ 1486 and AD Calvados, C 4226.
[34] Of all marsh drainage cases in Normandy, only one was decided in favor of a lord. It was brought by the Abbey of Troarn and concerned a marsh that was quite distant from any village. The abbey, unlike lay lords, maintained exceptional records. Finally, this marsh, like those drained in the seventeenth and eighteenth centuries, was particularly deep, so that villages could not claim pasture rights on most of its surface.
[35] See AN H¹ 1496.

$$d(1) = (p(1 - \beta_w) + (1 - p)(1 - B_s))E_l(c_l).$$

Had the opposite (British) rule applied, the probability that courts would have decided a marsh was waste would have been

$$d(0) = p\beta_w(1 - E_v(c_w)) + (1 - p)(\beta_s)(1 - E_v(c_s))$$
$$+ p(1 - \beta_w) + (1 - p)(1 - B_s).$$

As noted in Proposition 1, $d(0)$ represents a larger proportion of marshes drained than $d(1)$. Furthermore, if, as argued, evidence presented by parties determined only a fraction of all cases, the probability that a marsh would be deemed waste under the British rule would be significantly higher than under the French one. The conclusion seems inescapable: Had France had Britain's burden-of-proof rule, far more marshes would have been drained before the French Revolution.

The institutional regime that prevailed after the French Revolution was markedly different from the one that had held sway under the Old Regime. Under the new regime, drainage proceeded swiftly, indicating that there were multiple solutions to the institutional problems of the Old Regime. One could have assigned the property rights to the lord (as in Britain). One could also have given them to the village and thereby reduced the transaction costs surrounding intravillage bargaining. Under either solution to the institutional problem, drainage would have proceeded, which is the crucial issue in terms of economic growth. But these two possible solutions had dramatically different redistributional consequences. Under the Old Regime, political organizations seem to have been so concerned with redistribution that they preferred to stifle economic growth rather than resolve issues of uncertain property rights. During the French Revolution, marshes were allocated to villages and not to lords. While villages benefited from the transfer, what mattered most was that drainage could proceed.

The simple model that has been analyzed here helps to identify one of the many causes of the divergent experiences of England and France: differing burden-of-proof rules. The conclusion that legal rules have an important impact on economic growth goes far beyond the question of marsh drainage. Historians have noted that, while British lords and landowners seem to have retained ultimate control of the land, French peasants ultimately prevailed over lords to become landowners. Robert Brenner has suggested that the state played a crucial role in upholding the rights of peasants in France and those of lords in England.[36] This chapter has

---

[36] Brenner (1982).

argued that, in an epoch of highly uncertain property rights, a change in legal institutions such as the burden-of-proof rule is sufficient to alter the distribution of land ownership dramatically. Indeed, it can be shown that the difference between burden-of-proof rules that applied to marshes in England and in France also applied to the question of medieval peasant property. Thus, in France judicial rules favored peasant property, while in England lords were more likely to receive all residual claims to the land. If economic development required significant and costly coordination among landholders, the allocation of land to peasants rather than to lords may have slowed economic development.

Our exploration of the allocation of marshes between waste and common, and in the long run between drained and undrained land, highlights the importance of institutions in explaining economic performance. It seems inescapable that France had a set of institutions that were particularly hostile to drainage, whereas England's institutions encouraged lords to drain marshes.

10

Conclusion

A historian's opinion of the Revolution's economic consequences is frequently forged from the archival material he or she has to rely on. To a historian interested in trade, the dramatic fall in activity after 1789 in ports like Bordeaux and Marseille speaks of economic decay and Revolutionary failure. On the basis of such evidence, Alfred Cobban in *The Social Interpretation of the French Revolution* gave a scathing review of the economic consequences of 1789.[1] But for a historian who focuses on agriculture, the quiet that fell on the countryside after 1800 speaks of an end to the long-running tensions that had plagued the eighteenth century. To be sure, the post-1789 countryside was no Eden, but landowners were now able to go about their business unencumbered by the tangle of property rights that had been the hallmark of the Old Regime.

Cobban focused on the short run to understand whether the Revolution had improved the lot of those who had witnessed it. For the immediate post-Revolutionary period, his conclusions seem valid: "France was worse off in 1815 than she had been in 1789."[2] In this book, however, the focus has been on the long run. From this perspective, it seems just as clear that the French economy improved as a result of Revolutionary reform. Cobban's conclusion, after all, was derived from the implicit assumption that all the economic tragedies that unfolded between 1789 and 1815 were direct consequences of the Revolution. Yet any short-run evaluation of the Revolution is saddled with the problem of assigning the cost of the wars that raged between 1792 and 1815. If the Revolution is made responsible for the bellicose policies that prevailed for the quarter-century after 1792, then no cost–benefit analysis

---

[1] Cobban (1968). Two fundamental facts explain his position. First, he looked at France's economy during the Napoleonic period (1798–1815), and second, he used foreign trade and manufacturing as his key indicators. To be sure, the Napoleonic Wars did not bring internal stability to France, and the war with England clearly had a disastrous effect on international trade.

[2] Cobban (1968, 79); Postel-Vinay (1989).

can make the Revolution appear worthwhile. But if the wars had causes other than Revolutionary expansionism, the short-run conclusions are rather different.

The problem is that military expansion had also been a long-standing policy of the Old Regime; the Revolutionary regimes were not alone in waging war. Under the Old Regime, the pursuit of warfare had been hampered by limited resources. The monarchy had been constrained by its lack of control over taxes. The Revolution dramatically increased the French government's resources and thus its ability to wage war. It is less clear, however, that it marked a turning point in foreign policy. Whatever the domestic costs of the wars, the successes of French troops abroad allowed the governments of Revolutionary France to achieve – and surpass – the international goals of the absolutist monarchy. It can be argued that France's international adventures were not simply the consequences of the Revolution, but rather were policies shared by nearly all French governments between 1700 and 1815. If so, then economic and political reforms, rather than wars, were the primary consequences of the Revolution.

The investigation of property rights in agriculture confirms another of Cobban's assertions, that the beneficiaries of the French Revolution were "landowners, *rentiers,* and officials."[3] Indeed, the Revolution accomplished goals that were important to such individuals. The administration was strengthened and centralized, while property rights were clarified and secured. But those who controlled France after 1795 refrained from wholesale interference in the economy. Property rights were simplified by altering the balance of power in the countryside. Under the Old Regime, power at the local level had been divided between feudal authorities, landowners, and the rest of the community. The Revolution of 1789 eliminated seigniors and gave landowners control of village councils. After the Revolution, landowners alone controlled local agricultural policies. The policies they chose were far from ambitious: Between 1815 and 1850, no wholesale enclosures or consolidations occurred, and very little land already under plow was enhanced by drainage. Investments flowed mostly into improvements, like drainage and irrigation, that did not require altering the distribution of land. The Revolution was therefore not a sudden metamorphosis that transformed France into a modern society overnight. Rather, it was a crucial first step in reforming the property rights that hampered growth.

The reforms of the Revolution were first applied in secondary sectors like irrigation and drainage. As a result, the impact of Revolutionary reforms seems modest, because they initially benefited sectors that played a limited role in the overall growth process. Water control projects real-

---

[3] Cobban (1968, 173).

ized during the early part of the nineteenth century contributed to increased agricultural output by about 5 percent. The limited impact of irrigation and drainage in the early nineteenth century may have been largely technological. It was possible to irrigate only a small amount of land without dams and by gravity alone. Similarly, it was possible to drain only a limited amount of land without pumps or clay pipes. Nonetheless, after 1815, within the bounds imposed by technology, many improvements occurred in striking contrast to the Old Regime's overall failure.

Techniques like clay pipe drainage were applied with increasing frequency after 1860, partially because the state subsidized them and partially because their costs had fallen. Yet the property rights reforms of the Revolution were a crucial prerequisite for the application of these novel techniques. Under Old Regime property rights arrangements, the contractual problems that had prevented the reclamation of marginal lands were minor compared with those that promoters of drainage on land already under the plow would have faced. Furthermore, the development of the transportation system under the Old Regime would no doubt have been marred by the problems of irrigation development as well as by the prevailing system of internal tariffs. To overcome these problems and apply the techniques of the Industrial Revolution required property rights that were different from those of Old Regime France, property rights made possible by the political structure that emerged from 1789. To the extent that the Old Regime could have effectively centralized power and solved its public finance problems, the Revolution was unnecessary. Yet those two reforms threatened the very essence of Old Regime *privileges*. Before 1789, organizations like the judiciary blocked all reform to prevent a weakening of *privileges*. Thus, reform required a significant redistribution of political power.

The technological developments that diffused into French agriculture after 1860 were a long way off in 1789. Even though such new techniques contributed more to economic growth than did property rights reform, they were not available during the eighteenth century, nor in fact during the early nineteenth century. Old Regime administrators, well aware that they could not influence the pace of technological change, turned to schemes like water control to spur the economy forward. Although increased investment in irrigation and drainage depended on a government reform of property rights, the Old Regime failed to achieve any economically meaningful change. The success of the Revolution in agriculture is that it significantly improved institutions, including those that permitted long-run technical change. In this sense, the Revolution brought France into the modern world.

Conclusion

The cost of the institutions that constrained the French economy was well known under the Old Regime, and most of the reforms carried out after 1789 had been proposed long before. Yet the central government failed to implement reforms that would have promoted economic change. In fact, the institutional legacies of the Middle Ages and the early modern period were property rights arrangements that stymied development. Yet when such arrangements were chosen, they must have been somehow attractive. To understand why institutional change was short-sighted rather than efficient, let us briefly consider the history of property rights over wetlands.

The important institutions governing reclamation were defined in two phases, first in the Middle Ages and later during the early modern period. During the first phase, royal judicial officials forced lords, rather than villagers, to bear the burden of proof in disputes over landownership. The Crown chose a burden-of-proof rule that favored villagers because it aimed to reduce the power of nobles, not because it wanted to encourage wetland reclamation. Nonetheless, there may have been few immediate economic costs to a policy that favored villages and strengthened the king's hand in dealing with nobles.[4] In fact, there is no indication that it had much impact on agriculture before the seventeenth century. Drainage projects suffered only in the second phase, when the Crown sought to solve a second problem – an erosion of the tax base. A long-run transfer of land from villages into the hands of tax-exempt owners was eating away at the tax base, and by 1670 the king was compelled to prohibit sales of village land. This piece of legislation dramatically reduced the possibility of marsh drainage given that villages controlled most wetlands. Indeed, under the new rules marshland was common land, and any change in its use required the consent of all villagers. In the late seventeenth century, the cost of such rigidity was small because France faced a deep economic crisis and land was relatively abundant. Only after half a century of economic recovery did villages' control of marshes and the near impossibility of changing the status of wetlands collide with the further development of French agriculture.

In order to understand the institutions governing water control before the Revolution, we must understand how they too came to prevail. It appears that these institutions where attractive to the Crown and private individuals because they solved short-term problems. In the Middle Ages, kings, seigniors, and villages attempted to contract efficient allocations of resources without much consideration for the future. In fact, it seems that the state as well as individuals chose institutions that reduced im-

---

[4] The short-term costs of such a policy have little to do with wetlands; rather, they involve the existence of economies of scale in agriculture that would be realized if lords, rather than villages, retained property of the land. Cf. Brenner (1982).

174

portant transaction costs even at the expense of flexibility. For example, in the Middle Ages, access to marshes was closely regulated while ownership remained poorly defined. In the short run, uncertain property rights were actually efficient. Marshland reclamation, which required well-defined property rights, was unlikely because the demand for land was low. Regulating access without explicitly specifying transferable property rights reduced the necessity of costly specification of individual property rights to land of low value in the short run. Problems arose later when the demand for land increased and the improvement of marshes became potentially profitable. In the long run, uncertain property rights made reclaiming marshes difficult because no one had the authority to decide on drainage and everyone could claim rights to compensation.

Up to the Revolution, law making and private contracting were nearsighted for two reasons. First, the extremely low rate of economic change encouraged a static perspective among decision makers. No doubt it seemed unlikely to decision makers that marshes would become valuable in their lifetime.[5] Moreover, until the eighteenth century the economy was dominated by short-run price uncertainties due to the crises surrounding poor harvests and high mortality rather than secular change due to economic development. Thus, property rights may have been structured more to overcome short-term uncertainty than to accommodate long-term change.[6]

The second factor that encouraged short-term decisions was the shortsightedness of most actors. Medieval and even early modern governments had little interest in the distant future because they were most frequently engaged in a struggle for survival. This lack of foresight in government policy persisted into the eighteenth century. When considering a section of the economy as small as irrigation or as large as finance, the agents of Louis XIV could hardly have weighed the costs that would be imposed twenty years hence. What mattered to the Sun King was the battle for the mastery of Europe, not the future of canal building or even the development of capital markets.[7] But if rulers were ever changing, institutions were strikingly resilient. Thus, the cost of laws imposed by medieval and early modern rulers on irrigation and drainage proved to be quite large.

If institutions were primarily created in response to short-term considerations, they persisted because the short-term returns to change did not

---

[5] One can argue that the Black Death was an example of extremely rapid change, but it was to a large extent unanticipated. As a result, it would not have affected the decisions of individuals before 1340.

[6] A classic example of such property rights arrangements is the open-field system. See McCloskey (1975).

[7] In any case, one should not indict the Old Regime alone for a problem that has plagued governments throughout history. Indeed, while individuals and politicians may consider the impact of their actions on the future, they are concerned primarily with the short-term consequences of their actions.

compensate for the costs of reform. Medieval institutions, after all, were remarkably resistant to change – some were not modified until the Revolution. Given the renewed profitability of reclamation after 1730, however, one might expect that both individuals and the state would have altered institutions to take advantage of the new economic opportunities. Yet the fundamental barrier to institutional change appears to have been the refusals of individuals, groups, and the king to bear the redistributional consequences of reform. Had it been possible to enact property rights reforms free of redistributional consequences, change might have occurred under the Old Regime. The primary problem was the need to compensate for losses suffered as a result of reforms. Given any reform proposal, two types of equitable compensation plans were theoretically possible. One plan would have paid all claims to compensation, valid or otherwise. Since many fallacious claims would most likely be filed, such largesse would require, at the very least, that any change in property rights feature enormous benefits. However, few reform proposals had enough benefits to pay all claims, so outright compensation was not a viable plan. The other equitable plan would have determined the validity of each claim before making compensation. Since property rights were uncertain, determining the validity of claims for compensation would also have been very costly. Moreover, there could be no assurance that there existed sufficient evidence to determine the validity of claims. In fact, it was unlikely, even with a cooperative judiciary, that one could determine exactly the value of existing property rights. Therefore, the second type of plan was also doomed to failure. Without definite knowledge of the property rights structure, it was simply impossible to devise a fair reform plan, because valid compensation claims would not be distinguished from fallacious ones.

Reforms could have been undertaken by two broad classes of actors: private individuals or public organizations. Given the inefficiencies associated with the prevailing institutional structure, one might presume that individuals would have improved arrangements on their own. One requirement of private reform, however, is that it receive the approval of all concerned. Thus, all claims to compensation must be either paid or investigated, in either case a very costly procedure. One simple reason for the persistence of antiquated institutions was therefore that the very uncertainty of property rights made private institutional reform – when it was legal – economically unfeasible.

Since individuals were unable to resolve the property rights problems of the Old Regime, the state should have shouldered the burden. During the eighteenth century, the Crown made numerous attempts to reform "feudal" property rights. Yet the state faced systematic opposition to its efforts to achieve change, and its efforts came to naught. Individuals claimed – truthfully or not – that they would suffer from the proposed reforms, and they sued in courts that were all too willing to block royal initiatives. Opponents had an easy task because the Crown sought to impose reforms from above without full compensation for lost property rights.

Opposition to royal initiatives was fierce because change featured both redistributional consequences and opportunities for rent seeking. For example, common wetlands could have been given to villages or to their seigniorial lords, or divided between the two. Yet both sides had equally valid claims to all of the wetlands. Hence, any reform proposal could have been seen as redistributing from one group to the other, which guaranteed opposition from villages, from lords, or from both. In addition, opposition occurred simply because it might lead reformers to increase compensation plans for lost property rights. Both villagers and lords knew well that the Crown was willing to revise reforms in their favor in order to reduce opposition. Not surprisingly, until the Revolution placed marshes in the hands of villages, reform was fiercely opposed.

While the fate of reform attempts depended on the strength of local opposition, it also depended on the institutions and organizations that decided how laws should be changed. The Old Regime was saddled with a particularly costly mechanism for legal reform. In eighteenth-century France, the organizations and institutions that controlled change were as much a legacy of the past as the laws that controlled land and water. France was forged from a collection of autonomous regions where, until the late Middle Ages, different institutions had evolved. As the Crown asserted its authority over an increasing number of regions, it sought to avoid revolt by guaranteeing a certain amount of local autonomy. In part to make this promise credible, the Crown created specific organizations, *parlements,* to monitor royal legislation so that it would not conflict with existing regional institutions or *privileges.* Yet the Crown never fully accepted the *parlements'* oversight, and the king's agents often strove to increase their power at the expense of regional authorities. Because "feudal" property rights were part and parcel of local *privileges,* reforming property rights of any sort was a touchy political issue. The Crown's motives for reforming *privileges* were multiple. To be sure, the Crown wished to liberate the economy from certain burdens. But it also wanted to assert its authority – including the right to sell new *privileges* – over *parlements* and other local organizations. *Parlement* officers derived much of their status and income from the protection of *privileges,* so they opposed reform to protect their social and economic position. In the case of water control at least, they effectively blocked royal legislative efforts.

Within this political system, the Crown could choose two avenues for reform, contention or consensus. In theory, the Crown could override local organization, but the government was not prepared to bear the costs associated with despotism. Meanwhile, consensus, at least in the case of drainage and irrigation, was nearly impossible and certainly never achieved before 1789. The same was true of other areas.[8] The prevalence of opposition simply discouraged state initiative. As a result, agricultural reforms were proposed only when diverse conditions were satisfied. First, reform attempts required highly committed royal ministers for whom the rewards of success would have been large, financially or politically. Second, change occurred only at times when land prices were rising rapidly because this increased the social rewards of successful reforms. Third, reforms demanded relative state solvency so that the Crown could bear the inevitable costs of unpopular change: restricted access to credit and difficult tax collection.

Although the Revolution was not brought about because of the dissatisfaction with common land regulation, eminent domain authority, or water rights, these institutions were costly to the economy. It also seems hard to deny that, in many sectors of the French economy, the prevailing property rights arrangements reduced investment.[9] While the property rights in question did not all date back to the Middle Ages, most had been created to solve problems in economic environments that were radically different from those prevailing in the eighteenth century. While in manufacturing and trade these property rights may have only reduced the levels of output, in water control they prevented investment outright.

By the 1780s, a continuance of the Old Regime seemed to satisfy only a small minority. In fact, it can be argued that the crisis of 1789 was precipitated because the essence of absolutism, royal control of taxes, had become unacceptable to all. The demand for reform emanated not only from critics of the government, but also from administrators themselves. The Crown's agents wanted reform because they wanted to implement a rational and more productive tax system. The third estate wanted an end to the unfair tax system. The other two orders wanted a greater share of power. The problem was that institutional reforms required a redistribution of power, something inconsistent with the maintenance of absolutism. In some cases, like water control, reform required strengthening the government's hand, while other cases, like finance, required a weakening of royal power. The convening of the Estates General, itself anathema to absolutism, created a mechanism whereby power could be reallocated and reform could proceed.

The Revolution had nothing directly to do with drainage or irrigation. Most landowners simply watched it all happen from afar and tried to take advantage of the redistributional opportunities that change pre-

---

[8] For example, see the attempts to liberate the grain trade from tolls and tariffs. Progressive ministers like Turgot sought to reform the Old Regime, yet they seem to have misunderstood the political economy problem (Baker, 1975, 59–61). For a more general analysis of grain commerce, see Kaplan (1984).

[9] See, e.g., Bossenga (1988).

Conclusion

sented. The events of 1789 were less about economics than about power. Yet the distribution of political power that emerged from the Revolution allowed institutional reforms to go forward. The National Assembly was the first organization that could claim to speak for the nation. In practice, since the Assembly spoke for the nation, it could rule over the nation without conferring with other organizations. For the first time in France's history, an organization had the authority necessary to revise all contractual agreements that had bound the state to individuals and groups. The creation of the National Assembly invalidated the old political structure that had blocked change and destroyed the traditional alliance between holders of *privileges* and local organizations. The Assembly's political claims did not prevail without conflict, but unlike the monarchy, Revolutionary governments were prepared to bear the costs necessary to impose their policies.[10] The success of the Revolution meant that the National Assembly and all the governments that followed 1789 would hold much greater power than the last absolutist monarch. Because the government was powerful and well funded, it could credibly commit itself to simple plans for water control. Moreover, since the Revolution had dramatically clarified property rights over land and water, development could proceed without conflict.

After the heady days of 1789, when property rights were redistributed wholesale, a slow process of institutional refinement continued to simplify the task of water control projectors. Over the course of the nineteenth century, technological change in water control and the rise of a modern transportation system increased the returns to investment in irrigation and drainage. It is frequently argued that these later developments played important roles in the promotion of agricultural investment. Yet without the Revolution's clarification of common property rights, the institutional and technological innovations of the nineteenth century would have been of little use. The high degree of uncertainty in Old Regime property rights ensured that, in the absence of reform, conflicts over the ownership and control of land and water would no doubt have continued to monopolize the energies and resources of landowners. Because of the very uncertainty of property rights, however, reform could not have occurred without dramatic redistribution. Since redistribution of property was contingent on political change, it is impossible to separate the Revolution's economic reforms from the Revolution itself.

[10] For details, see Sutherland (1986).

Appendix 1: wages, land prices, and interest rates

The sources of the data used to construct price series in Provence and Normandy are similar. For land prices and interest rates, I relied on notarial archives, while for wages I relied on the accounts of public organizations. The French *notaires* have no proper equivalent in the Anglo-Saxon world. The *étude de notaire* was a form of recording office that was neither truly private nor truly public. It was private in that public officials could not request notarial records, and public in that offices were sold by the state. Private contracts had to be signed in front of, and deposited with, a *notaire* in order to ensure judicial enforcement in case of noncompliance. In the rural world, both in the eighteenth and nineteenth centuries, *notaires* copied down all sort of contracts that transferred property rights between individuals. The copies they kept were bound in chronological order irrespective of the nature of the act. As a result, loans, mortgages, land sales and rentals, wills, estate inventories, public auctions, and marriage contracts run one after the other in notarial minutes. Yet these records rarely offer much in the way of tables of contents except for nominative indexes. Thus, although notarial archives offer a wealth of information pertinent to economic history, the collection of data from such sources is painfully slow.

The sources of wages require less introduction. Most of the price data that are available for Europe come from similar sources. I relied on the account books of three types of institutions to collect wage estimates: religious organizations, hospitals, and cities. For the South, at least, municipal accounts are usually very detailed. They offer an abundant and relatively untapped source of information to carry through studies of changes in wages and prices. In Normandy, the archives of cities have suffered much both from the weather and from destruction during World War II; as a result, I relied on a greater variety of archival sources for this region. For greater clarity, I shall detail the construction of each series by region, considering first Normandy and then Provence.

Wages, land prices, and interest rates

Land prices

Although historians of Normandy have published some land rental data for the first half of the nineteenth century, there are no published land price series that span the period from 1700 to 1860.[1] Similarly, published wage series give information for parts of the eighteenth century and for the period after 1830.[2] But there were no data to splice them together. To overcome these difficulties, I constructed both a wage and a land price series from archival material, which were then checked for consistency with the published evidence.

I collected data from a sample of land sales and rental contracts and constructed a price series from notarial archives – the only homogeneous source for land sale prices in the eighteenth and nineteenth centuries. I used the archives of the notary in Troarn to collect data on all the land sales and leases for every fourth year from 1702 to 1870.[3] The data set represents 1,241 contracts over forty-one sample years. (The notarial archives are missing for 1778 and 1798.) Unfortunately, there were not enough contracts to allow the construction of separate arable land and pastureland time series. Nor could a time series be constructed for Norman marshland, because marshes were neither sold nor rented in a market context between 1700 and 1870. Although separate series for arable land and marshes would have been preferable, there are sufficient data to estimate the opportunity cost of specific marshes, and the average land price series can be used as a lower bound for the price of pasture, because pasture always commanded a higher price than arable land.[4]

Two methodological issues had to be confronted in order to use these data. First, heterogeneity in quality and locational advantages could easily account for the variance in the land prices. I chose to ignore the issue of heterogeneous land quality for a variety of reasons. Among these was the fact that there seemed to be no systematic changes over time in the location, or reported types, of land traded. On top of this, the contracts sampled concerned land in a small area – less than a ten-mile square – whose primary market crop was livestock. Although produced locally, grain was probably traded only locally; there is evidence, however, that in the eighteenth century cattle from the area were sold in a regional market and, after 1800, in a national market. Finally, the argument pre-

---

[1] Desert (1977, 780–7) presents data on rentals for the period 1813–1900. El Kordi (1970, 248–51) has some data on land prices, but they are not sufficiently detailed to be of use.

[2] Desert (1977, 776–9); El Kordi (1970, 256); Perrot (1975, 1039–41).

[3] AD Calvados, 8E 25117–353, Notary of Troarn.

[4] To convert Old Regime surface measures to metric units, I relied on Navel (1932, 3–218); and Desert (1962, 67–76).

sented in Chapter 7 does not rest on determining the timing of a relative price change within a four-year period. What we are interested in is determining a range of potential profits for a given period. Within each time period, the variability of land prices – and as a result profits – gives a measure of the risk involved in draining marshes.

The second methodological problem was aggregating leases and sales. This was dealt with in a similarly pragmatic fashion. Aggregating leases and land sales requires tackling the problem of choosing an interest rate to capitalize leases and accounting for anticipated capital gains. The true measure of the social value of an improvement project involves all the income it generates, that is, both the short-term increase in rental income from the improved land and the long-term increase in capital gains. Since the long-term trend of prices was positive, aggregating both capitalized leases and sales contracts creates a downward bias in the estimate of land prices. A low price for land can only increase the confidence of our profit estimates. To capitalize leases, I used the rate that prevailed in mortgage markets throughout the period (5 percent per annum). Alternative measures of interest rates had little effect on the price series (Table A1.1).

Wages

There are three sources of published data for wages. Mohamed El Kordi has published some data for Bayeux from 1673 to 1784, and Gilles Perrot has compiled wages for Caen for the late eighteenth century.[5] Finally, Gabriel Desert has produced several wage series from 1830 to 1900.[6] Yet these series could not be spliced together because there was an important gap: 1789–1830. To overcome this lacuna I relied on the account books of the Abbey of Troarn for the eighteenth century and the accounts of the hospital of Lisieux for the period 1794–1814.[7] I used Desert's data for the period 1830–70. For the period 1814–30, in the absence of alternative information, wages were linearly interpolated.

Wage data were collected for a wide variety of occupations. The sample contains observations on 672 wage bills for seventeen professions. The only ones for which there were sufficiently frequent observations to compute time series were the building trades (carpenters, roofers, and masons). Carpenters and roofers earned comparable wages while masons on average earned more. The series presented in Table A1.1 consists of the average wages for carpenters and roofers. Each observation was weighted by the number of days that were paid.

[5] El Kordi (1970, 256); Perrot (1975, 1039–41). [6] Desert (1977, 776–9).
[7] AD Calvados, H 7997–8 and H sup. Lisieux M 22–31. Other sources were combed for information: AD Calvados, H 64–5 (Abbaye d'Ardenne) and F 6910 (harvest wage bills).

Table A1.1. *Price series for Normandy and the government
interest rate*

Date	Daily wage[a] (unfed, francs)	Land price (francs per hectare)	Number of contracts[b]	Interest rate[c]
1702	0.58	445	27	—
1706	0.72	555	30	—
1710	0.72	586	13	15.0
1714	0.91	534	16	12.0
1718	0.91	694	51	7.1
1722	0.91	1,131	10	6.3
1726	0.90	838	12	3.3
1730	0.90	863	50	4.7
1734	0.80	355	7	4.1
1738	0.80	738	32	3.5
1742	0.80	1,150	29	4.0
1746	0.70	672	36	5.3
1750	0.80	835	28	3.7
1754	0.80	961	36	5.1
1758	0.80	1,199	17	5.1
1762	0.85	1,339	25	6.8
1766	0.85	999	10	6.0
1770	0.90	1,467	10	10.0
1774	1.00	1,680	24	7.0
1778	1.00			6.3
1782	1.10	1,940	6	6.0
1786	1.10	1,935	17	5.7
1790	1.30	2,503	5	6.0
1794	2.00	4,017	7	5.5
1798	—	—	—	—
1802	1.50	2,610	14	8.2
1806	1.20	1,004	12	9.0
1810	1.70	3,540	12	7.0
1814	2.00	2,669	2	11.0
1818	1.75	1,998	7	7.0
1822	1.70	3,505	13	6.4
1826	1.80	2,680	24	6.4
1830	1.75	3,321	18	5.1
1834	1.75	2,736	29	4.4
1838	1.75	2,275	43	4.0
1842	1.80	2,815	41	4.2
1846	1.95	2,199	135	3.6
1850	2.50	2,147	32	3.6
1854	2.25	3,119	46	4.2
1858	2.45	3,275	79	3.5
1862	2.50	3,287	36	4.0

Appendix 1

Table A1.1. *(cont.)*

Date	Daily wage[a] (unfed, francs)	Land price (francs per hectare)	Number of contracts[b]	Interest rate[c]
1866	2.50	3,260	41	4.8
1870	2.50	2,482	35	4.1

[a] Semiskilled workers (carpenters and roofers).
[b] Number of contracts used to form the land price estimate.
[c] The data before 1752 are estimated. See text and Table A1.2 for details.
Sources: For wages and land prices, see text. Nineteenth-century interest rate data from Homer (1977, 195–6). Eighteenth-century data from David Weir and François Velde, "The Financial Market and Government Debt in France, 1750–1793" (Yale Univ., mimeo, 1990).

Interest rates

Choosing an appropriate rate of interest to compare hypothetical rates of return proved to be more difficult than anticipated. Rental prices for land, from the notarial data I collected, ran at 5 percent of the sale prices throughout the period 1702–1870. This points to a real interest rate of 5 percent if we ignore appreciation in the value of the land. Mortgages point to the same stable rate of 5 percent. At the same time, however French interest rates in various capital markets fluctuated between 4.75 and 8.25 percent in the nineteenth century. David Weir and François Velde have collected data for the period 1752–92 which suggest that interest rates fell in a similar range both before and after the Revolution.[8]

Unfortunately, there are no private rates of interest for the eighteenth century except for *rentes* (personal loans), which ranged between 3 and 5 percent. All of this suggests that 5 percent was a reasonable upper bound for French interest rates in the eighteenth century and something closer to a lower bound in the nineteenth century. Choosing a 5 percent interest rate for the entire period from 1700 to 1848 seems a reasonable assumption, which will bias my test against the profitability of drainage projects before 1800. This can only strengthen my findings.

As an alternative, I estimated French interest rates from British data by running a regression of French rates for the period 1752–1870 on British interest rates for consols and a constant. I then used the British rates (which are available throughout the eighteenth century) to extrapolate French rates backward from 1750 to 1710. The regression results are

[8] David Weir and François Velde, "The Financial Market and Government Debt Policy in France, 1750–1793" (Yale Univ., mimeo, 1990). Eugene White made available to me similar data on tape.

Wages, land prices, and interest rates

Table A1.2. *Linear regression estimates of French interest rates*

Independent variable	Estimated coefficient	Standard error
Constant	−2.56	1.160
British rates	1.52	0.322
French rate 1[a]	0.89	0.081
French rate 2[a]	−0.41	0.082
Standard error of the regression		2.29
Durbin–Watson statistic		1.80
$N = 117$, $R^2 = .64$, $\bar{R}^2 = .63$, SSR[b] $= 595$		

Note: The dependent variable is French interest rates.

[a] French rate 1 and French rate 2 are the interest rates lagged one and two years to correct for serial correlation.

[b] SSR, Sum of squared residuals.

Sources: British interest rates and nineteenth-century French rates from Homer (1977, 156–7, 195–6, 222–3). Eighteenth-century French rates from David Weir and François Velde, "The Financial Market and Government Debt Policy in France, 1750–1793" (Yale Univ., mimeo, 1990) and from personal communications with Eugene White.

presented in Table A1.2. When available, actual French rates are always about 1 percent higher than the British rates. However, when British interest rates were very low, as in the middle of the eighteenth century, the extrapolation yields French rates that were too close to British rates and probably unreliable. When calculating rates of return to drainage projects, I used both the 5 percent rate of interest and the rate estimated from British data for my comparisons. The final interest rate series is presented in Table A1.1.

PROVENCE

Land prices

At least four notarial *études* (practices) were active in Cavaillon between 1700 and 1855. However, gathering data from all land contracts was impractical.[9] Sampling was therefore necessary. The data represent a complete, quinquennial sample of both land-sale and land-rental contracts from one *étude* from 1745 to 1855. Up to 1745 I sampled two

---

[9] A complete enumeration of land transactions for Cavaillon from 1700 to 1855 would have taken at least two years of research.

études because the first one had too few land contracts.[10] The total sample contains 1,781 observations. To obtain both an irrigated- and a dry-land price series, it was necessary to distinguish sales and rental of irrigated land. Before 1800, such sorting was relatively easy, since all the contracts contained detailed information about the quality of the land. After 1800, however, notaries ceased recording such information regularly. I therefore relied on location data to distinguish between irrigated and dry land after 1800.[11] Such sorting by location is imperfect, making the irrigated series a downward-biased estimate of irrigated land prices and the dry series an upward-biased estimate of the price of dry land after 1800. The bias is probably rather small considering that sorting the eighteenth-century data according to location or according to reported land quality yields very comparable estimates.

As in the case of Norman prices, Provençal land prices were estimated by a simple procedure that enabled me to use data from both rental contracts that predominated before 1789 and from sales contracts that predominated after the Revolution. For rental contracts, the value of a transaction was computed by capitalizing the rent using the interest rate. As noted earlier, such a procedure ignores capital gains; thus, the price series for dry and irrigated land is downward-biased. If anything, this reduces the absolute value of the difference in these prices, which brings down the estimated rates of return. For sales, the value of land was simply the price. For each year and for each type of land (dry or irrigated), average prices were calculated as the sum of the value of all transactions divided by the total area sold.

Wages

One excellent source of wage data is René Baehrel, *Une croissance: La Basse Provence rurale;* unfortunately, Baehrel's data stop in 1789.[12] There are no published data on Provençal wages for the nineteenth century. To overcome these difficulties, a sample of wage bills was collected that contained data on 851 bills of wages for thirteen professions, covering roughly 150 years for a total of more than sixty thousand man-days.[13] The bills were sorted into skilled and unskilled. The reported wages are the ratio

---

[10] AD Vaucluse, Fond Liffran, 659, 664, 669, 674, 679, 771, 775, 777, 779, 682; Fond Rousset, 904, 907, 909, 912, 915, 918, 912, 924, 926, 929, 932b, 935, 940, 945, 950, 955, 960, 965, 795, 975, 976, 977, 982, 987, 992, 997, 1002, 1007, 1012, 1017, 1022, 1029–30, 1047–8.

[11] To convert Old Regime surface measures, I relied on Barral (1876, chap. 6).

[12] Baehrel (1962, pp. 604–14).

[13] The sources include the account books of the city of Avignon (AC Avignon, CC 550 to CC 805, *pièces à l'appuis des Comptes*), religious institutions (AD Vaucluse, H Bompas 182–5; H Cordeliers d'Avignon, 62–4), and the hospital of Avignon (AD Vaucluse, H sup. Hôpital Ste Marthe E 103, M 6–18).

Table A1.3. *Price series for Provence, 1700–1855*

Date	Daily wage (unfed, francs)		Land price (francs per hectare)		Interest rate (percentage per year)
	Unskilled	Skilled	Dry	Irrigated	
1700	0.61	1.00	511	2,153	5.06
1705	0.77	0.90	814	2,730	5.15
1710	0.76	1.07	1,080	2,201	4.84
1715	0.74	1.00	851	2,819	4.96
1720	0.75	1.00	1,510	2,590	4.69
1725	0.74	1.12	939	2,297	5.45
1730	0.79	1.11	1,031	2,202	5.20
1735	0.76	1.22	1,598	2,407	4.80
1740	0.73	1.10	1,320	2,355	4.74
1745	0.92	1.25	1,538	2,477	5.16
1750	0.88	1.14	1,827	2,694	5.34
1755	0.94	1.25	1,827	2,694	5.09
1760	0.76	1.30	1,934	2,857	5.22
1765	0.93	1.25	1,410	2,867	5.04
1770	1.09	1.25	2,361	3,854	5.05
1775	1.12	1.60	2,081	3,055	5.03
1780	1.12	1.50	2,203	5,003	4.80
1785	1.18	1.50	1,604	4,328	5.10
1790	1.14	1.50	2,756	3,271	6.87
1795	—	—	4,365	6,309	—
1800	1.66	2.41	1,910	2,963	9.30
1805	1.15	2.24	2,044	5,438	8.70
1810	1.59	2.16	2,698	4,097	6.15
1815	1.61	2.43	2,640	3,831	7.40
1820	1.78	2.27	2,420	3,677	6.70
1825	1.74	2.22	2,985	5,100	5.09
1830	1.73	3.18	2,590	4,411	5.16
1835	1.89	3.00	2,450	4,469	4.33
1840	1.78	2.60	2,399	5,149	4.58
1845	1.77	3.00	2,621	5,192	3.62
1850	1.88	3.00	2,679	5,400	3.57
1855	1.86	3.00	2,900	5,470	4.41

Source: See the discussion in this appendix.

of the total wage bill for a given year divided by the total number of days worked. To the extent that canals were built solely in the winter – when wages were lower than average – the rates of return would be even higher than estimated. Yet since winter and summer wages seem to move in tandem, the trends in the wage bills should not depend on the aggregation procedure. Indeed, the distribution of wage bills over the year did not vary much over time. Since we care most about the trend in benefit –cost ratios, the procedure used here seems adequate.

Interest rates

For interest rates in Old Regime Provence, I relied on a sample of credit contracts from notarial archives. Detailed examination of this data set is part of another research project.[14] For the nineteenth century, I relied on the same data as for Normandy. Table A1.3 presents wages, land prices, and interest rates for Provence.

[14] See Jean-Laurent Rosenthal, "A Credit Market in Old-Regime France: l'Isle-sur-Sorgues, 1650–1788" (UCLA Dept. of Economics Working Paper 589).

Appendix 2: estimating rates of return

The construction of hypothetical benefit–cost ratios and internal rates of return for Normandy was given a great deal of attention in the text. Thus, there is no need to go over the construction of those series. The complete set of estimates is presented in Table A2.1. In the case of Provence, however, two different issues must be explored. Indeed, Chapter 7 left in the dark both how the estimate of increased irrigation was constructed and how canal construction accounts were put together.

INCREASES IN IRRIGATED AREAS IN PROVENCE

The studies of irrigation by J.-A. Barral offer good data for estimating the increase in total output as a result of irrigation because he investigated irrigated acreage thoroughly for each canal. According to Barral, in 1875 the total irrigated area for Provence was about 52,700 hectares, or 18 percent of the total cultivated area.[1] To evaluate total output changes, we must know the increase in area irrigated, not only from the Durance (a figure that is available and presented in Table A2.2), but also from other rivers. The increase in irrigation from other sources is not known precisely, so I present two estimates. The first concerns the impact of the increase in Durance irrigation, while the second estimates the increase in output that would have occurred if irrigation from all sources had grown at the same rate as Durance irrigation.

Since irrigation at least doubled agricultural output, we know that the increase in output will be close to the ratio of newly irrigated area to total cultivated area. The early-nineteenth-century canals added 16,314 hectares of irrigated land from the Durance compared with 3,211 for the

---

[1] One hundred forty thousand acres. The total cultivated area in the Bouches du Rhône and the Vaucluse was 201,000 hectares (excluding olive groves and vines). Since I was concerned with output that could be increased by irrigation, I excluded both olives and vines from my measure of total cultivated area. Barrral (1875, 323–4; 1875–6, Vol. 1, 83–7, 511–12).

Appendix 2

Table A2.1. *Hypothetical profits for Norman drainage projects*

| | Hypothetical annual rates of return (%) | | Hypothetical benefit–cost ratios | | | |
| | | | Government rate model | | Mortgage rate model | |
Date	Terriers	Troarn	Terriers	Troarn	Terriers	Troarn
1702	26.3	11.7	—	—	1.6	1.0
1706	31.6	12.7	—	—	1.8	1.0
1710	33.3	24.6	2.4	1.3	1.9	1.1
1714	23.5	−38.5	1.7	0.9	1.4	0.8
1718	37.0	10.4	2.1	1.1	1.9	1.0
1722	58.7	144	3.2	1.7	3.1	1.6
1726	44.9	57.2	2.4	1.3	2.3	1.2
1730	46.6	64.9	2.5	1.3	2.4	1.3
1734	8.46	−78.2	1.1	0.6	1.0	0.6
1738	41.8	54	2.2	1.2	2.2	1.2
1742	63.4	198	3.5	1.9	3.4	1.9
1746	40.4	65	2.3	1.3	2.2	1.3
1750	47.1	88	2.5	1.4	2.5	1.4
1754	53.8	132	2.9	1.6	2.9	1.6
1758	64.7	215	3.6	2.0	3.6	2.0
1762	68.9	236	4.4	2.4	3.9	2.1
1766	54.6	125	3.1	1.7	2.9	1.6
1770	72.4	251	5.3	2.8	4.1	2.2
1774	76.4	265	5.0	2.6	4.3	2.2
1778	—	—	—	—	—	—
1782	81.8	289	5.1	2.5	4.7	2.3
1786	81.8	287	4.9	2.5	4.6	2.3
1790	90.5	334	5.8	2.8	5.3	2.6
1794	104.0	357	6.3	2.8	6.0	2.7
1798	—	—	—	—	—	—
1802	88.7	282	5.9	2.8	4.9	2.3
1806	45.6	31	2.8	1.4	2.2	1.1
1810	103.0	377	6.9	3.2	6.1	2.8
1814	80.2	169	5.0	2.3	4.0	1.8
1818	68.9	116	3.8	1.7	3.3	1.5
1822	103.1	371	6.7	3.1	6.0	2.8
1826	84.5	212	4.8	2.2	4.4	2.0
1830	98.0	326	5.6	2.6	5.6	2.5
1834	82.8	219	4.3	2.0	4.6	2.1
1838	70.4	134	3.4	1.5	3.8	1.7
1842	83.5	214	4.2	1.9	4.6	2.1
1846	61.3	72	2.8	1.2	3.3	1.5
1850	51.5	−2	2.2	0.9	2.6	1.1
1854	75.4	164	3.9	1.7	4.2	1.8

Table A2.1. *(cont.)*

| | Hypothetical annual rates of return (%) | | Hypothetical benefit–cost ratios | | | |
| | | | Government rate model | | Mortgage rate model | |
Date	Terriers	Troarn	Terriers	Troarn	Terriers	Troarn
1858	75.4	126	3.4	1.5	4.1	1.8
1862	78.6	138	3.6	1.6	4.1	1.7
1866	82.8	157	4.0	1.7	4.0	1.7
1870	62.6	52	2.8	1.2	3.1	1.3

Sources: See Appendix 1 and Chapter 6.

Table A2.2. *Increases in area irrigated from the Durance (hectares)*

| | Period | | | |
Region	1100–1700	1700–89	1790–1820	1820–60
Vaucluse	3,835	2,253	0	9,105
Bouches du Rhône	10,624	1,765	0	8,211
Total	14,459	4,108	0	16,316

Sources: See this appendix and Barral (1876, 323–34; 1875–6, Vol. 1, pp. 86–91, 511–12).

eighteenth century as a whole. The Durance's increase alone would have led to a 7.7 percent increase in total output for Provence. If non-Durance irrigation had witnessed the same growth, total output would have increased by about 12 percent.

Construction accounts for Provence

Canal construction accounts rarely itemized costs beyond excavation *(terrassement)* and skilled construction *(ouvrages d'arts)*. As the former was done by unskilled labor, I divided those costs by the wage for unskilled labor for the period in which the project was carried out to get an estimate of the quantity of labor employed. To simplify the calculation of the rates of return, I assigned all skilled construction and administrative costs to skilled labor. Skilled construction involved the building of bridges for roads over canals and aqueducts for canals over small rivers and valleys. Such jobs were clearly the domain of skilled masons. Non-labor inputs were also assigned to skilled labor, because the primary input of canals other than labor was quarried stone. Quarrying was an extractive industry that required only skilled labor and some transportation. Thus, the cost of quarried stone should have closely followed the price of skilled labor. The sum of nonlabor inputs, skilled labor, and administrative expenses was divided by the skilled wage to get an estimate of the number of skilled man-days.

Land consumed by canals

The main canal of Carpentras, the largest canal in my sample, was only 7.5 meters wide. Including the embankments, it occupied an area less than 17 meters across for the first quarter of its length. The rest of the main canal occupied an area less than 10 meters wide, and its branches were even smaller. Other canals were less than 4 meters at their widest and their branches were much smaller than that. I assumed that all canals required a band of land 15 meters wide. This band had a length equal to the length of the canal and its main branches. This simplifying assumption reduced the estimated profits and thus strengthened my findings that irrigation was profitable before 1760.[2]

Uncounted revenues

I disregarded certain revenues accruing to canals that were difficult to estimate. These revenues came from the sale of water power rights on canals to mill owners. To be sure, mills were an important source of revenues for some canals. For example, they brought in revenues equal to one-sixth of the maintenance costs on the canal of Crillon. A mill was worth more than 20,000 livres in the eighteenth century, or more than 5 percent of the cost of a small canal.[3] The size and value of mills varied greatly. Moreover, the value of a mill is not a good indicator of the rent paid by a mill owner for a waterfall, which is what accrued to the canal owner. Therefore, one would need not only the rental contracts of mills but also their agreement with canal owners to ascertain what mill operators paid for falls. The effort to secure rental contracts would thus be very great for little gain. Obviously, the omission will push my hypothetical rates of return downward.

[2] Caillet (1925, Vol. 2, 194–212).
[3] AD Vaucluse, 1 doc. 221. See also AD Vaucluse, S Usines et Cours d'Eaux (Avignon, canal of Crillon, 1820).

Table A2.3. *Hypothetical benefit–cost ratios for irrigation projects*

Date	Cabedan-Neuf I	Cabedan-Neuf II	Crillon	Plan-Oriental	Carpentras
1700	3.85	3.20	3.66	3.82	3.13
1705	4.01	3.45	3.91	4.34	3.33
1710	2.06	1.83	2.06	2.27	1.75
1715	3.96	3.42	3.88	4.24	3.30
1720	1.91	1.80	1.99	2.29	1.71
1725	2.57	2.24	2.53	2.74	2.17
1730	2.14	1.88	2.12	2.32	1.82
1735	1.29	1.20	1.34	1.49	1.14
1740	1.82	1.67	1.87	2.07	1.59
1745	1.08	0.92	1.04	1.23	0.88
1750	1.30	1.23	1.36	1.58	1.16
1755	1.22	1.14	1.27	1.45	1.07
1760	1.33	1.28	1.42	1.60	1.20
1765	2.24	2.02	2.26	2.53	1.93
1770	1.96	1.92	2.11	2.48	1.78
1775	1.15	1.06	1.18	1.33	1.01
1780	3.31	3.10	3.44	3.93	2.92
1785	3.43	3.04	3.42	3.82	2.92
1790	0.56	0.55	0.60	0.70	0.51
1795	—	—	—	—	—
1800	0.87	0.75	0.85	0.92	0.70
1805	3.02	2.65	3.00	3.27	2.49
1810	1.19	1.09	1.22	1.37	1.02
1815	0.95	0.85	0.96	1.06	0.80
1820	1.03	0.91	1.03	1.14	0.86
1825	1.71	1.57	1.75	1.99	1.49
1830	1.29	1.14	1.28	1.37	1.10
1835	1.46	1.27	1.44	1.55	1.24
1840	2.18	1.92	2.16	2.36	1.86
1845	1.88	1.67	1.87	2.02	1.62
1850	1.92	1.69	1.91	2.07	1.63
1855	1.81	1.62	1.82	1.98	1.57

Sources: Tables 7.1 and 7.2 and this appendix.

Maintenance costs

Some maintenance costs already appear in the price of irrigated land. Indeed, the price of a particular piece of irrigated land is equal to the discounted stream of profits from using that land minus the capitalized value of whatever maintenance costs are assessed on that land. If all irrigated land were assessed uniformly, there would be no need to count maintenance costs, but such uniformity was far from prevalent in Ca-

Table A2.4. *Hypothetical internal rates of return for irrigation projects*
(percentage per year)

Date	Cabedan-Neuf I	Cabedan-Neuf II	Crillon	Plan-Oriental	Carpentras
1700	157.5	84.6	120.4	156.1	50.1
1705	166.0	102.4	127.2	185.0	54.1
1710	77.9	35.4	68.8	94.3	30.1
1715	157.3	93.2	121.0	173.0	51.5
1720	61.3	27.9	58.2	86.5	25.8
1725	95.5	45.0	80.5	105.0	34.8
1730	77.3	32.4	67.6	89.4	29.2
1735	25.3	−3.7	27.6	41.3	10.2
1740	55.6	19.0	52.3	71.3	22.7
1745	35.5	3.1	35.8	50.9	14.4
1750	26.7	−0.8	29.6	47.5	11.3
1755	21.4	−6.7	24.1	38.8	8.3
1760	32.6	5.99	35.7	57.5	14.7
1765	88.5	47.1	77.6	111.0	34.4
1770	64.0	35.9	62.1	98.4	28.7
1775	17.1	−11.1	19.6	32.6	5.8
1780	127.1	85.3	106.7	164.2	48.0
1785	138.5	84.7	111.2	161.0	48.5
1790	−36.3	−53.4	−44.2	−24.0	31.2
1795	—	—	—	—	—
1800	−6.86	−36.5	−12.1	−0.9	10.8
1805	93.5	39.8	77.8	98.6	32.8
1810	12.7	−17.4	13.4	23.1	2.4
1815	−7.8	−10.6	−10.2	15.3	−8.3
1820	17.6	−12.5	18.9	29.9	5.3
1825	58.4	24.8	55.7	81.5	24.6
1830	49.7	15.2	47.9	67.8	20.4
1835	58.9	21.1	54.6	75.4	23.5
1840	89.3	44.7	77.3	107.4	33.9
1845	59.2	17.7	53.6	67.4	22.9
1850	66.6	24.2	59.6	77.1	25.7
1855	61.4	21.9	56.1	76.2	24.1

Sources: Tables 7.1 and 7.2 and this appendix.

vaillon, where each canal had a different organization dealing with main-tenance. Each organization assessed landowners on a yearly basis for contributions, but they did not assess land uniformly, either over time or across parcels. Thus, the land price series only reflects maintenance costs as assessed by the institutions governing canals in Cavaillon. It is clearly wrong to assume that the maintenance costs already affecting the irri-gated price series are the correct ones for all projects. As a result, it seemed best to assume that the price series reflects the discounted future revenues from land and to account for maintenance costs explicitly. To do this, and to simplify the calculation of internal rates of return, I assumed that the promoters created a sinking fund to pay for future maintenance costs. The complete series of hypothetical profits appears in Tables A2.3 and A2.4.

Appendix 3: theoretical proofs

The proof of Theorem 1 uses backward induction. First, I solve for the optimal expenditures c_w^*, c_s^*, and c_l^* given q and S. Second, I solve for the optimal litigation probabilities k_w^*, k_s^* given S. Third, I solve for the optimal settlement offer S^*.

Legal expenditures

Assume that the settlement offer S has been refused and that the postsettlement probability that the village is weak is q. The players now simultaneously choose their legal expenditures. The village chooses c_i to maximize $\Lambda_i(c_i)$. The first-order condition is

$$\frac{\partial \Lambda_i}{\partial c_i} = \frac{\partial E_v(c_i)}{\partial c_i} (\alpha - P)(1 - t)\beta_i - 1 = 0.$$

The lord must choose c_l to minimize $\Pi_l(c_l)$. The first-order condition is

$$\frac{\partial \Pi_l}{\partial c_l} = \frac{\partial E_l(c_l)}{\partial c_l} (\pi - \alpha)t(q(1 - \beta_w) + (1 - q)(1 - \beta_s)) - 1 = 0.$$

The concavity of $E_v(\cdot)$ and $E_l(\cdot)$ as well as assumption (5) of Chapter 9 are sufficient to guarantee that a solution to both problems will exist. First-order conditions will be necessary and sufficient because concavity ensures that the second-order conditions hold. Thus c_i^* and c_l^* are defined by

$$\frac{\partial E_v(c_i)}{\partial c_i} = \frac{1}{(\alpha - P)\beta_i(1 - t)},$$

$$\frac{\partial E_l(c_l)}{\partial c_l} = \frac{1}{(\pi - \alpha)t(q(1 - \beta_w) + (1 - q)(1 - \beta_s))}.$$

Theoretical proofs

Note that only the lord's litigation expenditures depend on litigation probabilities (k_i) through q.

Optimal litigation levels

Fix S. Let $c_i^*(k_w, k_s)$, c_w^*, c_s^* be the equilibrium expenditures. Villages of type i will choose k_i^* to minimize Λ_i. Let x_j be the probability that villages of type j litigate. Then let $k_i^*(S, x_j)$ be the best response of villages of the other type to such a litigation probability. Since there are only two types of villages, we can write simply $k_i^*(S, x)$. Without loss of generality, we assume that $k_s^*(0, x) = 1$ (otherwise $S = 0$, which leads to a trivial problem). Furthermore, always accepting offers of P is a dominant strategy, because trials are costly, so $k_w(P, x) = k_s(P, x) = 1$. Throughout the proof of the next two lemmas, the reference to S will be suppressed because S will be fixed. Hence, $k_i(S, x)$ will simply be $k_i(x)$.

Lemma 1. Weak villages never litigate more than strong villages.

Proof. Suppose first that $k_s^* = 0$; then $s \geq \Lambda_s$, but $\Lambda_s > \Lambda_w$, so $S > \Lambda_w$ and $k_w^* = 0$. Now suppose that $0 < k_s^* < 1$; then $S = \Lambda_s$, but $\Lambda_s > \Lambda_w$, so $S > \Lambda_w$ and $k_w^* = 0$. Thus, weak villages will litigate with positive probability only if strong villages never settle. Therefore, weak villages never litigate more than strong ones. □

Let us now examine the litigation decision of strong villages. Going back to equation (4) of Chapter 9 and differentiating with respect to k_s gives

$$\frac{\partial V_s}{\partial k_s} = P - S + \Lambda_s(c_s^*) + k_s(1 - \beta_s) \frac{\partial E_l(c_l^*)}{\partial c_l} \frac{\partial c_l^*}{\partial k_s} t(\alpha - P) = 0. \quad (3)$$

Clearly

$$k_s(1 - \beta_s) \frac{\partial E_l(c_l^*)}{\partial c_l} \frac{\partial c_l^*}{\partial k_s} t(\alpha - P)$$

is the only part of (3) that depends on k_s, and it is nonnegative and increasing in k_s. Therefore, we need only evaluate $\partial V_s / \partial k_s$ at $k_s = 0$. If $\partial V_s / \partial k_s$ is negative at 0, then strong villages will settle with probability 1; if $\partial V_s / \partial k_s$ is positive, strong villages will always sue. The litigation of strong villages depends in part on what weak villages have decided to do:

$$\frac{\partial V_s}{\partial k_w} = k_s(1 - \beta) t(\alpha - P) \frac{\partial E_l(c_l^*)}{\partial c_l} \frac{\partial c_l^*}{\partial k_w} < 0. \quad (4)$$

Indeed, as k_w increases, the value of going to court falls for strong villages because the lord's belief about the likelihood that he faces weak

197

types increases, and he spends more on research. As the value of going to court falls, strong villages may sue less. It can be shown that $\partial^2 V_s / \partial k_s \, \partial k_w$ evaluated at $k_s = 0$ is negative for all values of k_w. Thus, three cases arise: (i) Strong villages settle independent of the decisions of weak villages; $\partial V_s / \partial k_w$ is negative at $k_w = 1$. (ii) Strong villages sue independent of the decisions of weak villages; $\partial V_s / \partial k_w$ is positive at $k_w = 0$. (iii) Strong villages settle as long as weak villages sue enough; in case (iii), there exists a $\bar{k}_w$ such that, if k_w is less than or equal to $\bar{k}_w$, then $k_s = 1$, and if k_w is greater than $\bar{k}_w$, then $k_s = 0$.

The same analysis also enables us to characterize the reaction function of weak villages. The profits of weak villages, however, increase with the probability that strong villages refuse the settlement offer. Indeed, when strong villages go to court with greater probability, lords fight less ($\partial^2 \Pi_l / \partial k_w^2 < 0$). Clearly, changes in type j's probability of going to court affect only the other type's profits of litigation, not those of accepting the lord's offer. The best response function of weak villages to the litigation probability of strong ones is monotonic, increasing, and continuous. The reaction functions of strong villages to the trial probabilities of weak villages is either constant at 1 or at 0, or for some $\bar{k}_w$ the profits of strong villages are maximized exactly when they refuse the settlement with probability 1 ($\partial V_s(1, c_s) / \partial k_s = 0$), in which case strong villages have a discontinuous reaction function. If the reaction function of strong villages is constant (with respect to k_w), it is trivial to show that there is a unique equilibrium pair of litigation probabilities k_s^*, k_w^*. However, the case remains open if strong villages have a discontinuous reaction function.

Lemma 2. If there is a discontinuity in k_s^* at $\bar{k}_w$, then $k_w^*(1) < \bar{k}_w$.

Proof. Given that the settlement offer is S and that weak villages litigate with probability $\bar{k}_w$, strong villages are at an optimum at $k_s^* = 1$, so $\partial V_s / \partial k_s$ is 0 when $k_w = \bar{k}_w$ and $k_s = 1$. Now let us look at $\partial V_w / \partial k$ evaluated at $k_w = \bar{k}_w$ and $k_s = 1$:

$$\frac{\partial V_w(\bar{k}_w, 1)}{\partial k_w} = P - S + \Lambda_w(c_w^*) + \bar{k}_w(1 - \beta_w) \frac{\partial E_l(c_l^*)}{\partial c_l} \frac{\partial c_l^*}{\partial k_w} t(\alpha - P). \quad (5)$$

$\Lambda_w(c_w^*)$ is less than $\Lambda_s(c_s^*)$, and

$$\bar{k}_w(1 - \beta_w) \frac{\partial E_l(c_l^*)}{\partial c_l} \frac{\partial c_l^*}{\partial k_w} t(\alpha - P)$$

is negative. Thus,

$$\frac{\partial V_w(\bar{k}_w, 1)}{\partial k_w} < \frac{\partial V_s(\bar{k}_w, 1)}{\partial k_s} = 0 \Rightarrow k_w^*(1) < \bar{k}_w. \quad \square$$

Lemma 2 guarantees the existence of a unique pair of equilibrium litigation probabilities despite the fact that the best response function of strong villages may not be continuous. Indeed, if their best response function is continuous, it is constant either at 0 or at 1; and then the fact that weak villages have a monotone best response function guarantees that the equilibrium will be unique.

If the best response function of strong villages is discontinuous, then it is 1 when k_w is low and 0 when k_w is high. Lemma 2 says that, in that case, strong villages will litigate with probability 1 and that weak villages will always accept the settlement offer. Thus, there is a single equilibrium pair of litigation probabilities for each offer S. Furthermore, strong villages always use pure strategies.

Definition. z_i is a reservation settlement for a village of type i if and only if (i) $k_i^* = 1$ for all $S > z_i$ and (ii) $k_i^* = 0$ for all $S < z_i$. z_i is a reservation-like offer if only (i) or (ii) holds.

Lemma 3. There exists one reservation settlement offer for strong villages and two reservation-like offers for weak villages.

Proof. The best response functions of all villages are decreasing with S:

$$\frac{\partial^2 V_i}{\partial S \, \partial k_i} = -1.$$

Given that $k_s^* = 0$ if $S = P$ and $k_s^* = 1$ if $S = 0$ and that k_s^* is a razor's edge reaction function, there must exist a z_s such that $k_s^* = 0$ if and only if $S > z_s$ and $k_s^* = 1$ if $S < z_s$. z_s is the reservation offer of strong villages. Similarly, define $\bar{z}_w$ such that $k_w^*(S) = 0$ if and only if $S > \bar{z}_w$. Define $\underline{z}_w$ as the S such that $k_w^*(S) = 1$ if and only if $S < \underline{z}_w$. $\underline{z}_w$ and $\bar{z}_w$ exist because the best response of weak villages is continuous and decreasing in S.$\square$

From Lemma 1 we know that the reservation offer of strong types is accepted by weak types with probability 1 ($\bar{z}_w < z_s$). Thus, if strong villages accept the settlement offer, then weak villages accept the offer as well. For any settlement offer above $\bar{z}_w$, $q = 0$ and the lord will spend the least documenting his claim. Define $\bar{c}_l$ to be his litigation expenditures in this case. Then $z_s = \Lambda_s(c_s^*, \bar{c}_l)$. Lemma 2 enables us easily to define out-of-equilibrium beliefs for the lord. The natural extension of q above z_s when no trials should occur is clearly $q = 0$. To find $\underline{z}_w$ and $\bar{z}_w$ it suffices to solve the first-order condition for $V_w(k_w, c_w)$, the profit function for weak villages:

$$S = \Lambda_w(c_w^*) + k_w(1 - \beta_w) \frac{\partial E_l(c_l^*)}{\partial c_l} \frac{\partial c_l^*}{\partial k_w} t(\alpha - P) + P. \tag{6}$$

$\underline{z}_w$ is the S that solves (6) when $k_w = 1$ and $k_s = 1$. $\bar{z}_w$ solves (6) when $k_w = 0$ and $k_s = 1$. To avoid the trivial equilibrium in which the settlement

Appendix 3

Table A3.1. *Optimal litigation probabilities given village type and settlement offer*

Offer	$S < \underline{z}_w$	$S \in [\underline{z}_w, \bar{z}_w]$	$S \in [\bar{z}_w, z_s]$	$S \geq z$
Weak village response (k_w^*)	1	[0, 1]	0	0
Strong village response (k_s^*)	1	1	1	0

offer is 0 and everyone accepts, I assumed that strong villages would always sue if offered nothing. However, weak villages may accept a 0 offer with positive probability ($\underline{z}_w \leq 0$ or $\bar{z}_w \leq 0$). If either $\underline{z}_w \leq 0$ or $\bar{z} \leq 0$, we can redefine them without loss of generality to be 0. The possible equilibrium litigation levels are described in Table A3.1.

Optimal settlement

Now that optimal expenditures and litigation probabilities have been determined, let us consider the choice of settlement offer by the lord.

Claim. There are only three potential equilibrium settlement offers: 0, $\bar{S}$, z_s, where $\bar{S}$ minimizes $(1 - k_w^*)(\pi - S) + k_w^* \psi(c_i^*)$ subject to $S \in [\underline{z}_w; \bar{z}_w]$.

Proof. All other potential equilibrium settlement offers are dominated by those just listed. For example, settlement offers greater than $\bar{z}_w$ but less than z_s are dominated by $\bar{S}$. If offered a settlement between $\bar{z}_w$ and z_s, weak villages still accept the offer with probability 1 but get a higher settlement. Strong villages still sue with probability 1. So the lord pays strictly more for no gain, and thus $\bar{z}_w$ dominates S, but $\bar{S}$ dominates $\bar{z}_w$. Now for $S \in [0, \underline{z}_w]$, 0 is a weak local minimum. For $S \in [\underline{z}_w; \bar{z}_w]$, the lord's revenue function is

$$\Pi_l(S, c_l) = p[1 - k_w^*](\pi - S) + k_w^*(\psi_w(c_i^*)) + (1 - p)(\psi_s(c_i^*)). \quad (7)$$

The first-order condition for the lord is

$$\frac{\partial \Pi_l(S, c_l)}{\partial S} = p[1 - k_w^*] - \frac{\partial k_w^*}{\partial S}(\psi_w(c_i^*) - S). \quad (8)$$

Note that $\partial k_w^*/\partial S$ does not depend on S. The second-order condition always holds because $\partial \Pi_l(S, c_l)/\partial S = -2 \, \partial k_w^*/\partial S > 0$. The loss function thus has a unique minimum between $\underline{z}_w$ and $\bar{z}_w$. Call $\bar{S}$ the settlement that maximizes the lord's profit between $\underline{z}_w$ and $\bar{z}_w$.

The only possible equilibrium offers are 0, $\bar{S}$, z_s. However, 0 is never an equilibrium offer unless $\bar{S} = 0$. The argument must be divided into two cases: $0 < \underline{z}_w$ and $0 > \underline{z}_w$. Consider first $0 < \underline{z}_w$. Now let us compute the difference in the revenues of the lord if he offers 0 and if he offers $\bar{S}$:

$$\Pi_l(0) - \Pi_l(\bar{S}) = p \, k_w^*(\psi_w(c_l^*) = \bar{S}). \tag{9}$$

But

$$\bar{S} = \Lambda_w(c_w^*) + k_w^*(1 - \beta_w) \frac{\partial E_l(c_l^*)}{\partial c_l} \frac{\partial_l^*}{\partial k_w} t(\alpha - P). \tag{10}$$

Note that

$$k_w^*(1 - \beta_w) \frac{\partial E_l(c_l^*)}{\partial c} \frac{\partial c_l^*}{\partial k} t(\alpha - P) < 0.$$

So $\Pi_l(0) - \Pi_l(\bar{S}) > (c_l^* + c_w + F_l + F_w)k_w^*$, which is positive. Therefore, 0 is dominated by $\bar{S}$.

Now suppose $0 > \underline{z}_w$; then $0 \in [\underline{z}_w; \bar{z}_w]$. On that interval, the unique minimum of the lord's revenue function is $\bar{S}$, so 0 cannot be an equilibrium offer unless it is $\bar{S}$. Therefore, there remain only two possible equilibria:

$$E_1 = \left(c_l^*, c_w^*, c_s^*, k_w^*(\bar{S}), 1, \bar{S}, \frac{pk^*}{pk^*(1-p)} \right),$$

$$E_2 = (\bar{c}_l, c_w^*, c_s^*, 0, 0, z_s, 0).^1$$

This completes the proof of Theorem 1. □

PROOF OF COROLLARY 1

Suppose the equilibrium is E_1; then $\Pi_l(\bar{S}) > \Pi_l(z_s)$. Let $\Pi_l(\bar{S}) - \Pi_l(z_s) = 2\varepsilon$. $\Pi_l(\cdot, x)$ is continuous in x, where x belongs to $\{\alpha, P, \pi, t, \beta_w, \beta_s, p, F_l, F_v\}$. Continuity ensures that for each x there exist a δ such that $|\Pi_l(\bar{S}, x + \gamma)| - |\Pi_l(\bar{S}, x)| < \varepsilon$ and $|\Pi_l(z_s, x + \gamma)| - |\Pi_l(z_s, x)| < \varepsilon$ for all $\gamma < \delta$. Thus, $\Pi_l(\bar{S}, x + \gamma) > \Pi_l(z_s, x + \gamma)$ for all $\gamma < \delta$. Now let δ^* be the smallest of all δ. E_1 is the unique equilibrium in the ball of radius δ^* centered at $\{\alpha, P, \pi, t, \beta_w, \beta_s, p, F_l, F_v\}$. Therefore, the equilibrium is locally unique.

---

[1] One should note that, strictly speaking, in E_2 strong villages are indifferent between settling and litigating. The indifference of strong villages suggests that there is a continuum of mixed strategies for these villages in this equilibrium and two pure strategies, one in which they sue and one in which they settle. Yet if strong villages sue with positive probability, then z_s no longer minimizes the lord's revenue function. So strong villages litigating with positive probability when offered z_s cannot be part of a sequential equilibrium. Thus, E_2 remains the only sequential equilibrium.

Bibliography

MANUSCRIPT SOURCES

Abbreviations in parentheses appear in the notes.

National Archives (AN)

Série D XIII 1
Série F 10, 208–9, 311–24
Série F 12, 1513–15
Série F 14, 127, 141–2, 627–30, 6300, 6391, 11168
Série F II, 910
Série H¹, 1226, 1260–1, 1303, 1307–10, 1387, 1486–1500, 1511–15, 1625–
   7, 1692

Calvados (Caen)

1. Departmental archives (AD)

2. Provincial and departmental series
Série C 4073–8, 4192, 4197–205, 4210, 4214–15, 4226–30, 4239–55, 4258,
   4260, 4262–7, 4270–77, 4288, 4293–8, 6771
Série F 6910
Série H 64–5 (Abbaye d'Ardenne), 7997–8, 8160–6 (Abbaye de Troarn)
Série H supp. Lisieux M 22–31
Série S 998, 1004, 1067–8, 1267–73, 2270, 5271

3. Communal archives (AC)
Amfreville 9 E 009/40
Bavent 9 E 046/69, 9 E 046/109 (Robehome)
Bures 563 E dt/15
Janville 9 E 344/46
Petiville 9 E 724/49 (Varavile)
Ranville 9 E 530/81–9
Saint Samson 9 E 657/33
Toufreville 9 E 698/48–50, Etang, Communaux
Vimont 9 E 761/75, Etang, Communaux (Saint Pierre Oursin)

4. Notarial archives
The notarial archives for Troarn are at the departmental archives in Caen. I sam-

Bibliography

pled the archives of the *notaire* at Troarn for land-sale and lease data every four years from 1702 to 1870. One year (1798) was missing. The call numbers ranged from 8 E 25117 (1702) to 8 E 25365 (1870).

Vaucluse (Avignon)

1. Departmental archives (AD)
Série C 34–46, 61
Série E, Fond Bressy du Thor 1–14
Série 4 G 2–5
Série H Celestin d'Avigon 57–9, Chartreux de Bompas 150, 182–5
  Cordeliers d'Avignon 37–9, Hospital Ste. Marthe 162–7
Série 1 L 365–78
Série S 6, Usines et Cours d'eaux. This series was under classification at the time I saw it. Documents had until then been assigned to boxes on the basis of the village of origin and which project they concerned. Thanks to the amiable cooperation of the staff at the archives, I was able to consult all the documents relevant to Cavaillon, Avignon, and l'Isle sur Sorgues, three towns with many canals and water wheels. This series is precious because it holds much evidence on the working of the nineteenth-century administration.

2. Printed documents
1 doc 220–5

3. Communal archives (AC)
Cavaillon, BB 20–4 archives of the Bishop (available in Cavaillon)
Avignon, CC *pièces à l'appuis des comptes* (account records) every five years from 1700 to 1790 (available at the departmental archives in Avignon)

4. Notarial archives
The notarial archives for Cavaillon have been given to the departmental archives in Avignon. Cavaillon always had at least four *notaires* between 1700 and 1860. I sampled the Blaze archives (fond Rousset) once every five years from 1700 to 1860. Because the Blaze archives had few land contracts before 1750, I also sampled the Lieutard archives (fond Liffran) from 1700 to 1745. The call numbers are Rousset 905–1048 and Liffran 659–729.

5. Mediathèque Cecano (BM) (Avignon Municipal library)
Manuscripts: 1558–60, 1605, 1632–5, 2060, 2433–6 2459, 2511–13, 2549, 2575, 2741, 2826, 2852, 2931, 2932, 2933, 2955, 3334.
Bound volumes of printed documents: 4°6198, 4°6824

Bouches du Rhône (Aix en Provence)

1. Communal archives (AC)
Série AA. 34–46

2. Bibliothèque Municipale (BM) Méjanes (Aix en Provence Municipal library)
Manuscripts 722(609), 741(614), 831(846), 834(848), 837(851), 840(853)
Bound volumes of printed documents: F 335, F 415

The primary sources of data I have not used in this study were the Archives des Bouches du Rhône because of a lack of time and resources. While there are many data on institutions and irrigation in these archives, the wealth of printed data as well as the information available in manuscript form in the Vaucluse did not make it necessary to use them.

Bibliography

PUBLISHED SOURCES

Alguhon, Maurice. 1970. *La vie sociale en Provence interieure au lendemain de la Révolution.* Paris: Société des Études Robespierristes.

Allen, Robert. 1982. The efficiency and redistributional consequences of eighteenth-century enclosures. *Economic Journal* 92:937–53.

——— 1983. Recent developments in production cost and index number theory, with an application to international differences in the cost and efficiency of steel making in 1907/9. *Historish Sozialwissenschaftliche Forchungen des Zentrum fur Historische Sozialforschungen* 15:90–99.

——— 1988. The growth of labor productivity in early-modern English agriculture. *Explorations in Economic History* 25:117–46.

——— Forthcoming. *Enclosures and the yeomen.* Oxford Univ. Press.

Allen, Robert, and Cormac O'Grada. 1988. On the road again with Arthur Young: English, Irish, and French agriculture during the industrial revolution. *Journal of Economic History* 48:93–116.

Arthur, Brian. 1989. Competing technologies, increasing returns, and lock-in by historical small events. *Economic Journal* 99:116–31.

Ault, Warren. 1972. *Open field farming in medieval England.* London: Allen & Unwin.

Baehrel, René. 1962. *Une croissance: La Basse Provence rurale de la fin du XVIe siècle à 1789.* Paris: SEVPEN.

Baker, Keith. 1975. *Condorcet from natural philosophy to social mathematics.* Chicago: Univ. of Chicago Press.

Barral, Jean-Auguste. 1876. *Les irrigations dans le département de Vaucluse.* Paris: Imprimerie Nationale.

——— 1875–6. *Les irrigations dans le département des Bouches du Rhône.* 2 vols. Paris: Imprimerie Nationale.

Barratier, Edouard. 1969. *Histoire de la Provence.* Toulouse: Privat.

Bates, Robert. 1981. *Markets and states in tropical Africa.* Berkeley and Los Angeles: Univ. of California Press.

Baulant, Micheline. 1968. Le prix des grains à Paris de 1431 à 1789. *Annales E.S.C.* 23:520–40.

——— 1971. Le salaire des ouvriers du bâtiment à Paris de 1400 à 1726. *Annales E.S.C.* 26:463–83.

Bebchuck, Lucian. 1984. Litigation and settlement under imperfect information. *Rand Journal of Economics* 15:404–15.

Beik, William. 1985. *Absolutism and society in seventeenth-century France: State power and provincial aristocracy in Languedoc.* Cambridge Univ. Press.

Bergeron, Louis. 1972. *L'épisode Napoléonien: Aspects interieurs.* Paris: Editions du Seuil.

Bertin, J. B., and P. Audier. 1904. *Adam de Craponne et son canal.* Paris: Champion.

Bien, David. 1987. Offices, corps and a system of state credit: The uses of privilege under the Ancien Régime. In Keith Michael Baker, ed., *The political culture of the Old Regime.* 2 vols. Oxford: Pergamon, 1:87–114.

Billioud, Joseph. 1956. Le vignoble Marseillais, du XIIIeme siècle à l'aduction d'eau de 1840. *Provence Historique, Mélanges Busquet* 6:166–85.

Bloch, Marc. 1929. La lutte pour l'individualisme agraire dans la France du XVIIIe siècle. *Annales d'Histoire Économique et Sociale* 2:239–83, 511–56.

——— 1966. *French rural history.* Berkeley and Los Angeles: Univ. of California Press.

Bibliography

Block, Maurice. 1856. *Dictionnaire de l'administration Française*. Paris: Berger-Levrault.

Blume, Larry, David Rubinfeld, and Paul Shapiro. 1984. The taking of land: When should compensation be paid? *Quarterly Journal of Economics* 99:71–92.

Bossenga, Gail. 1988. La Révolution Française et les corporations: Trois examples Lillois. *Annales E.S.C.* 43:405–26.

Braudel, Fernand, and C. E. Labrousse, eds. 1970. *Histoire économique et sociale de la France*, vols. 2, 3/1, and 3/2. Paris: Presses Universitaires de France.

Brenner, Robert. 1976. Agrarian class structure and economic development in pre-industrial Europe. *Past and Present* 70:30–75.

——— 1982. The agrarian root of European capitalism. *Past and Present* 97:16–113.

Brewer, John. 1988. *The sinews of power*. New York: Knopf.

Bulow, Jeremy. 1982. Durable goods monopolists. *Journal of Political Economy* 90:314–32.

Caillet, Roger. 1925. *Le Canal de Carpentras*. Carpentras: Imprimerie Batailler.

Chabert, André. 1945–9. *Essais sur le mouvement des revenus et de l'activité économique en France de 1789 à 1820*. 2 vols. Paris: De Medicis.

Chaline, Jean-Pierre. 1966. Les biens des hospices de Rouen: Recherches sur les fermages Normands du XVIIIᵉ au XXᵉ siècle. *Revue d'Histoire Economique et Sociale* 44:265–78.

Chandler, Alfred. 1977. *The visible hand: The managerial revolution in American business*. Cambridge, Mass.: Harvard Univ. Press.

Chaussinand-Nogaret, Guy. 1976. *Les financiers du Languedoc au XVIIIᵉ siècle*. Paris: SEVPEN.

——— 1985. *The French nobility in the eighteenth century*. Cambridge Univ. Press.

Ciriacono, Salvatore. 1989. Venise et la Hollande, pays de l'eau (XVᵉ–XVIIIᵉ siècles). In J.-L. Miege, M. Perney, and Ch. Villain-Gandossi, eds., *L'Eau et la culture populaire en Méditerranée*. Aix-en-Provence, 99–114.

——— 1990. Financing land reclamation in the 17th and 18th centuries: Towards a European model? In *Communications of the Tenth International Economic History Congress*. Lewen.

Clère, Jean-Jacques. 1988. *Les paysans de la Haute-Marne et la Révolution Francaise*. Paris: Éditions du C.T.H.S.

Clout, Hugh. 1983. *The land of France*. London: Allen & Unwin.

Coase, Ronald. 1937. The theory of the firm. *Economica* n.s., 4:386–405.

——— 1960. The problem of social cost. *Journal of Law and Economics* 3:1–44.

——— 1972. Durability and monopoly. *Journal of Law and Economics* 15:143–9.

Cobban, Alfred. 1968. *The social interpretation of the French Revolution*. Cambridge Univ. Press.

Cooke, George. 1864. *Acts of common in England and Wales*. London: Stevens.

David, Paul. 1985. Clio and the economics of QWERTY. *American Economic Review* 75:332–7.

Darby, H. C. 1983. *The changing fenland*. Cambridge Univ. Press.

Davis, Lance, and Douglass North. 1971. *Institutional change and American economic growth*. Cambridge Univ. Press.

Delalande. 1777. *Des canaux de navigation*. Paris: Dessaint.

Demsetz, Harold. 1964. The exchange and enforcement of property rights. *Journal of Law and Economics* 7:10–26.

——— 1966. Some aspects of property rights. *Journal of Law and Economics* 9:61–70.

Derlange, Michel. 1987. *Les communautés d'habitants en Provence au dernier siècle de l'Ancien Régime*. Toulouse: Le Mirail.

Bibliography

Desert, Gabriel. 1962. Mesures agraires anciennes et nouvelles dans le Calvados. *Annales de Normandie* 12:68–76.

———. 1977. *Une société rurale au XIXᵉ siècle: Les paysans du Calvados.* New York: Arno Press.

Dewald, Jonathan. 1987. *Pont Saint-Pierre, 1398–1789.* Berkeley and Los Angeles: Univ. of California Press.

Dienne, Louis de. 1891. *Histoire des desséchements des lac et marais en France avant 1789.* Paris: Champion.

Dion, Roger. 1934. *Le val de Loire, étude de géographie régionale.* Tours, Arrault.

———. 1961. *Histoire des levées de la Loire.* Paris.

Doyle, William. 1980. *Origins of the French Revolution.* New York: Oxford Univ. Press.

Duby, George. 1974. *The early growth of the European economy.* Ithaca, N.Y.: Cornell Univ. Press,

Duby, George, ed., 1975. *Histoire de La France rurale.* 3 vols. Editions du Seuill.

Durand, Yves. 1966. Recherches sur les salaires des maçons à Paris au XVIIIᵉ siècle. *Revue d'Histoire Economique et Sociale* 44:468–80.

Elie, Hubert. 1953. La spéculation sous la Régence: l'Affaire du canal d'Avignon à la mer. *Provence Historique* 3:112–30.

El Kordi, Mohamed. 1970. *Bayeux aux XVIIᵉ et XVIIIᵉ siècles.* Paris: Mouton.

Esmonin, Edmond. 1913. *La taille en Normandie au temps de Colbert (1661–1683).* Paris: Hachette.

Fenoaltea, Stephano. 1976. Risk, transaction costs, and the organization of medieval agriculture. *Explorations in Economic History* 13:129–51.

Fornery, Joseph. 1903. *Histoire du Comtat Venaissin.* Avignon: Seguin et Roumanille.

Forster, Robert. 1960. *The nobility of Toulouse in the eighteenth century: A social and economic study.* Baltimore: Johns Hopkins Univ. Press.

———. 1961. The Noble Wine Producers of the Bordelais in the Eighteenth Century. *Economic History Review* 14:18–33.

———. 1980. *Merchants, landords and magistrates: The Depont family in the eighteenth century.* Baltimore: Johns Hopkins Univ. Press.

Fruit, René. 1963. *La croissance économique du Pays de Saint-Amand (Nord) (1688–1914).* Paris: Armand Colin.

Furet, François, and Denis Richet. 1970. *The French Revolution.* New York: Macmillan.

Gangneux, Gérard. 1982. *L'ordre de Malte en Camargue du 17ᵉ au 18ᵉ siècle.* Grenoble.

Garnier, Bernard. 1979. Structure et conjoncture de la rente foncière dans le Haut-Maine au XVIIᵉ et XVIIIᵉ siècles. In Gabriel Desert et al., eds., *Problèmes agraires et société rurale.* Caen: Cahier des Annales de Normandie, No. 11:103–26.

Gerbaux, François, and Charles Schmitt. 1906–10. *Procès verbaux des comités d'agriculture et de commerce de la Constituante, de la Legislative et de la Convention.* 4 vols. Paris: Imprimerie Nationale.

Gilligan, Thomas, William Marshall, and Barry Weingast. 1989. Regulation and the theory of legislative choice: The Interstate Commerce Act of 1887. *Journal of Law and Economics* 31:35–61.

Goubert, Pierre. 1960. *Beauvais et le Beauvaisis de 1600 à 1730.* 2d ed. 2 vols.

———. 1982. Paris: Editions de l'Ecole des Hautes Etudes en Sciences Sociales.

———. 1973. *L'Ancien Régime.* Paris: Armand Colin.

Bibliography

Goubert, Pierre, and Denis, M. 1964. *1789: Les Français ont la parole* . . . Paris: Juliard.

Grantham, George. 1978. The diffusion of the new husbandry in Northern France, 1815–1840. *Journal of Economic History* 38:311–27.

1980.The persistence of open field farming in nineteenth-century France. *Journal of Economic History* 40:515–30.

1989. Agricultural supply during the industrial revolution: French evidence and European implications. *Journal of Economic History* 49:43–72.

Greif, Avner. 1989. Reputation and coalitions in medieval trade: Evidence on the Maghribi traders. *Journal of Economic History* 49:857–82.

Grenier, Jean-Yves. 1985. *Séries économiques Française: XVIᵉ–XVIIIᵉ siècles*. Paris: Editions de l'Ecole des Hautes Etudes en Sciences Sociales.

Gul, Faruk, Hugo Sonnenschein, and Robert Wilson. 1986. Foundations of dynamic monopoly and the Coase conjecture. *Journal of Economic Theory* 39:155–90.

Hammond, John, and Barbara Hammond. 1948. *The village labourer*. 2 vols. London: British Publisher's Guild.

Hauser, Henri. 1936. *Recherches sur l'histoire des prix en France*. Paris: Les Presses Modernes.

Head-König, Anne-Lise. 1972. Rente foncière et dîmes dans le Lyonnais au XVIIᵉ et XVIIIᵉ siècles: Leurs concordances. In E. Le Roy Ladurie and J. Goy, eds., *Les fluctuations du produit de la dîme*. Paris: Mouton, 153–65.

Hoffman, Philip. 1986. Taxes and agrarian life in early-modern France: Land sales, 1550–1730. *Journal of Economic History* 46:36–55.

1988. Institutions and agriculture in Old-Regime France. *Politics and Society* 16:241–64.

Forthcoming. Leases and agricultural productivity: The Paris basin, 1450–1789. *Journal of Economic History*.

Holtman, Robert. 1967. *The Napoleonic revolution*. Baton Rouge: Louisiana State Univ. Press.

Homer, Samuel. 1977. *A history of interest rates*. New Brunswick, N.J.: Rutgers Univ. Press.

Hyslop, Beatrice. 1933. *Répertoire critique des cahiers de doléances*. Paris: Leroux.

1936. *The general cahiers of 1789, with the texts of unedited cahiers*. New York: Columbia Univ. Press.

Isambert, Jourdan, and Decrusy, eds. 1821–33. *Recueil général des anciennes lois Françaises depuis l'an 420 jusqu'à la Révolution de 1789*. 29 vols. Paris: Belin-le-Prieur.

Jones P. M. 1988. *The peasantry in the French Revolution*. Cambridge Univ. Press.

Kahn, Charles. 1986. The durable goods monopolist and time consistency with increasing costs. *Econometrica* 54:275–94.

Kaplan, Steve. 1984. *Provisioning Paris, merchants and millers in the grain and flour trade during the eighteenth century*. Ithaca, N.Y.: Cornell Univ. Press.

Kreps, David. 1990. *A course in microeconomic theory*. Princeton, N.J.: Princeton Univ. Press.

Labrousse, C. E. 1933. *Esquisse du mouvement des prix et des revenus en France au XVIIIᵉ siècle*. 3 vols. Paris: Dalloz.

1943. *La crise de l'économie Française à la fin de l'Ancien Régime et au debut de la Révolution*. Paris: Presses Universitaires de France.

Lebrousse, C. E., Ruggiero Romano, and F.-G. Dreyfus. 1970. *Le prix du froment en France, 1726–1913*. Paris: SEVPEN.

Bibliography

Laffarge, René. 1902. *L'Agriculture en Limousin au XVIIIe siècle et l'intendance de Turgot*. Paris: Marescq.

Landes, David. 1969. *The unbound Prometheus*. Cambridge Univ. Press.

Le Roy Ladurie, Emmanuel. 1966. *Les paysans du Languedoc*. 2 vols. Paris: SEVPEN.

Lefebvre, Georges. 1959. *Les paysans du Nord*. Abridged ed. Bari: Laeterza.

1962. *Etudes Orléanaises*. 2 vols. Paris: Commission d'Histoire Economique et Sociale de la Révolution.

Libecap, Gary, and Steven Wiggins. 1985. The influence of private contractual failure on regulation: The case of oil field unitization. *Journal of Political Economy* 93:689–714.

Lindley, Keith. 1982. *Fenland riots and the English Revolution*. London: Heinemann.

Maidlow, John. 1897. *Six essays on common preservation*. London: Low.

Maistre, André. 1968 *Le canal des deux mers, canal royal du Languedoc, 1666–1810*. Toulouse: Privat.

Marczewski, Jean, ed. 1961. *L'Histoire quantitative de l'économie Française*. Paris: Cahier de l'Institut de statistique et d'économie appliquée, no. 115.

Martel, André. 1955. Les origines du canal de Plan Oriental. *Actes du Congres des Societés Savantes*, 385–405.

Masson, Paul. 1901. Le Canal de Provence. In *Revue Historique de Provence*: Aix en Provence, 350–9, 421–37.

1929–30. *Encyclopédie des Bouches du Rhône*. Vols. 4, 5, and 7. Paris: Champion.

McCloskey, Donald. 1975. The persistence of English common fields, and The economics of enclosure: A market analysis. Both in W. Parker and E. Jones, eds., *European peasants and their markets*. Princeton, N.J.: Princeton Univ. Press, 73–114 and 115–34, respectively.

Michel, Louis. 1978. Quelques données sur le mouvement de la rente foncière en Anjou, du milieu du XVIIe siècle à la Révolution. In E. Le Roy Ladurie and J. Goy, eds., *Prestations paysannes, dîmes, rente foncière et mouvement de la production agricole à l'époque préindustrielle*. 2 vols. Paris: Mouton, 2:607–24.

Merlin, M. 1828. *Questions de droit*. 4th. ed. 18 vols. Paris: Tarlier.

Meurer, Michael. 1989. The settlement of patent litigation. *Rand Journal of Economics* 20:77–91.

Meuvret, Jean. 1977–88. *Le problème des subsistences à l'époque Louis XIV*. 3 vols. Paris: Mouton.

Morineau, Michel. 1970. *Les faux semblants d'un démarrage économique: Agriculture et démographie en France au XVIIIeme siècle*. Cahier des Annales no. 30. Paris: Armand Colin.

Mousnier, Roland. 1971. *La venalité des offices sous Henri IV et Louis XIII*. Paris: Presses Universitaires de France.

1979. *The institutions of France under the absolute monarchy*. 2 vols. Chicago: Univ. of Chicago Press.

Navel, Henri. 1932. Recherches sur les anciennes mesures agraires Normandes, acres, vergées, perches. In *Bulletin de la Société des Antiquaires de Normandie*. Caen: Jouanet et Bigot.

Norberg, Kathryn. 1988. Dividing up the commons: Institutional change in rural France, 1789–99. *Politics and Society* 16:265–86.

North, Douglass. 1981. *Structure and change in economic history*. New York: Norton.

Bibliography

North, Douglass, and Barry Weingast. 1989. Constitution and commitment: The evolution of institutions governing public choice in seventeenth-century England. *Journal of Economic History* 49:803–32.

——— 1990. *Institutions, institutional change and economic performance.* Cambridge Univ. Press.

O'Brien, Patrick, and Caglar Keyder. 1978. *Economic growth in Britain and France, 1780–1914: Two paths to the twentieth century.* London: Allen & Unwin.

Ollivier and Sallembert. 1856. *Projet de desséchement de la Dives.* Caen.

Olson, Mancur. 1971. *The logic of collective action, public goods and theory of groups.* Cambridge, Mass.: Harvard Univ. Press.

Pelzman, Sam. 1976. Toward a more general theory of regulation. *Journal of Law and Economics* 19:211–44.

Perrot, Jean-Claude. 1975. *La genèse d'une ville moderne: Caen au XVIIIe siècle.* Paris: Mouton.

Petot, Jean. 1958. *Histoire de l'administration des ponts et chaussées de 1599 à 1815.* Paris: Rivière.

Pillorget, René. 1975. *Les mouvements insurrectionels de Provence entre 1596 et 1715.* Paris: Editions A. Pédone.

Poitrineau, Abel. 1979. *La vie rurale en Basse-Auvergne* (1st ed. 1965). Marseille: Laffitte reprints.

Ponteil, Fernand. 1965. *Les institutions de la France de 1814 à 1870.* Paris: Presses Universitaires de France.

Popkin, Samuel. 1979. *The rational peasant: The political economy of rural society in Vietnam.* Berkeley and Los Angeles: Univ. of California Press.

Postel-Vinay, Gilles. 1985. Les domaines nobles et le recours au crédit (France premier tiers du XIXe siecle). In *Les noblesses Européenes au XIXe siècle.* Rome, 199–220.

——— 1989. A la recherche de la révolution économique dans les campagnes (1789–1815). *Revue Economique* 40:1015–45.

Poterlet, T. 1807. *Code des desséchements.* Paris: Crozet.

Price, Roger. 1980. *The economic modernisation of France (1730–1780).* London: Croom Helm.

Priest, George, and Benjamin Klein. 1984. The selection of disputes for litigation. *Journal of Legal Studies* 13:1–55.

Rambert, Gaston. 1963. Le commerce de l'eau de vie à Toulon au XVIIIe siècle. *Provence Historique* 13:31–53.

Reboulet, A. 1914. Construction du canal de Crillon. *Mémoires de l'Academie du Vaucluse:* Avignon, 37–50.

Reinganum, Jennifer, and Louis Wilde. 1986. Settlement, litigation, and the allocation of litigation costs. *Rand Journal of Economics* 17:557–66.

Rigaud, Jean. 1934. *Le canal de Craponne: Etude historique et juridique relative aux concessions complexes des arrosages communaux d'Istre et Grans.* Aix en Provence.

Riou, René. 1987. *Les marais desséchés du Bas-Poitou* (1st ed. 1907) Marseille: Laffite Reprints.

Robb, Raphael. 1986. The Coase theorem: An informational perspective. Working Paper, Institute for Mathematics and Its Applications, Univ. of Minnesota.

Roehl, Richard. 1976. French industrialization: A reconsideration. *Explorations in Economic History* 13:233–81.

Rothwell, Harry. 1974. *English historical documents, 1189–1337* (Vol. 3 of En-

Bibliography

glish historical documents, ed. David Douglas). London: Eyre & Spottis-woode.

Rubinfeld, David, and D. E. M. Sappington. 1987. Efficient awards and standards of proof in judicial proceedings. *Rand Journal of Economics*, 18:308–15.

Saint-Jacob, Pierre. 1960. *Les paysans de la Bourgogne du Nord au dernier siècle de l'Ancien Régime*. Paris: Les Belles Lettres.

Sée, Henry. 1906. *Les classes rurales en Bretagne du XVIe siècle à la Révolution*. Paris: Briere.

Servais, Paul. 1984. *La rente dans le ban d'Herve*. Brussels: Crédit Communal de Belgique.

Schama, Simon. 1989. *Citizens: A chronicle of the French Revolution*. New York: Knopf.

Simiand, François. 1932. *Le salaire, l'évolution sociale et la monnaie: Éssai de théorie experimentale du salaire*, 3 vols. Paris: F. Alean.

Sion, Jules. 1909. *Les paysans de la Normandie Orientale*. Paris: Armand Colin.

Sobel, Joel. 1985. Disclosure of evidence and resolution of disputes: Who should bear the burden of proof? In A. E. Roth, ed., *Game theoretic models of bargaining*. Cambridge Univ. Press.

Spencer, William, and Cornelius Gillam. 1952. *A textbook on law and business*. New York: McGraw-Hill.

Stein, Robert. 1979. *The French slave trade*. Madison: Univ. of Wisconsin Press.

Stigler, George. 1971. The theory of economic regulation. *Bell Journal of Economics and Management Science* 3:3–21.

Stokey, Nancy. 1982. Rational expectations and durable goods pricing. *Bell Journal of Economics and Management Science* 12:112–28.

Stone, Lawrence. 1965. *The crisis of the aristocracy, 1558–1641*. New York: Oxford Univ. Press.

Sueur, Philipe. 1982. *Le conseil provincial d'Artois, 1640–1790*. 2 vols. Arras: Mémoire de la Commission Départementale des Monuments Historiques de Pas de Calais.

Summers, Dorothy. 1976. *The great level*. London: David & Charles.

Sutherland, D. M. G. 1986. *France, 1789–1815: revolution and counterrevolution*. New York: Oxford Univ. Press.

Syndicat du Canal de Cabedan-Neuf. 1883. *Archives et documents, 1230–1883*. Cavaillon: Imprimerie Mistral.

Taylor, George. 1964. Types of capitalism in eighteenth-century France. *Economic History Review* 79:478–97.

Non capitalist wealth and the origins of the French Revolution. *American History Review* 72:469–96.

Thirsk, Joan. 1967. *The agricultural history of England and Wales*. Vol. 4. Cambridge Univ. Press.

Tullock, Gordon. 1967. The welfare costs of tariffs, monopolies and theft. *Western Journal of Economics* 5:224–232.

Umbeck, John. 1981. *A theory of property rights with application to the California gold rush*. Ames: Univ. of Iowa Press.

Vidalenc, Jean. 1952. *Le département de l'Eure sous la monarchie constitutionelle, 1814–1848*. Paris: Rivière.

Villeneuve, Jean de. 1825–9. *Encyclopédie des Bouches du Rhône*. Vols. 2, 4, and 5. Marseille: Ricard.

Vovelle, Michel. 1972. *La chute de la monarchie, 1787–1792*. Paris: Editions de Seuil.

Bibliography

Weir, David. 1989. Tontines, public finance, and revolution in France and England, 1688–1789. *Journal of Economic History* 49:95–124.

———. Forthcoming. "Economic Crises and the Origins of the French Revolution." *Annales E.S.C.*

White, Eugene. 1989. Was there a solution to the Ancien Regime's financial dilemma? *Journal of Economic History* 49:545–67.

Wiggins, Steven, and Gary Libecap. 1985. Oil field unitization: Contractual failure in the presence of imperfect information. *American Economic Review* 75:370–85.

Young, Arthur. 1929. *Travels in France during the years 1787, 1788, and 1789.* Cambridge Univ. Press.

Zinc, Anne. 1969. *Azereix: La vie d'une communauté rurale à la fin du XVIIIe siècle.* Paris: SEVPEN.

Zolla, David. 1893. Les variations du revenu et du prix des terres en France au XVIIe et au XVIIIe siècles. *Annales des Sciences Politiques* 8:299–326, 439–61, 686–705.

Index

abbeys, 40
agriculture: development of, 13–14, 55, 57, 70; French compared with British, 12–14, 149–50; institutions in, 13, 91–3; policies toward, 52–3; productivity of, 101–2
Allen, Robert, 11, 15
Anjou, 65, 67–8
Apostolic Chamber, 112
Arles, 41, 65
Arthur, Brian, 33
Artois, 50–1
Assemblée du Pays, 113, 117, 119
associations (for rural improvement), 53–4, 58, 97–8, 146
Aubepagne canal, 44
Aunis, 42
Auvergne, 43
d'Avenel, [?], 91
Avignon, 116

Baehrel, René, 186
Barneville, 89
Barral, Jean-Auguste, 189
Bebchuck, Lucian, 155
Bertin, Henri, 92
Bien, David, 147
Blossac, [?] de, 88
bogs (peat), 16–17, 43
Boisgelin canal, 117
Bordeaux, 43, 171
Bouches du Rhône, 56, 111
Bouillon, duke of, 140
Boullonmoranges, 93–4, 139–40
Bourgoin, lords of, 132
Bourgoin marsh, 132, 134
Bradley, Humphrey, 41–4
Brenner, Robert, 168
Bulow, Jeremy, 131
burden-of-proof rules, 154, 158–9, 162, 174; in England, 163–4; in France, 165–9

Bures, 93
Burgundy, 51

Cabedan-Neuf canal, 104, 115–16, 120
Cabourg, 73
Caen, 83, 93, 99
Calvados, 56, 67–8, 71, 98
Cambis, duke of, 114
Cambis canal, 114
canals, 108, 146
Carpentras canal, 104, 120
Cavaillon, 65, 100–1, 104, 115–16, 133–4, 138–9
Cavaillon, bishop of, 133, 138–9, 141
Champenois, Noel, 42
Chandler, Alfred, 22, 26
Chateaurenard canal, 116n
Clout, Hugh, 55
coalitions, 28
Coase, Ronald, 22–3, 126, 131
Cobban, Alfred, 9, 171
common land, 15–16; division of, 32–3, 48, 50–1, 52, 96–8, 151–3; ideology of, 17; legal status of, 45–7, 93; litigation over, 139–40; rights to, 21, 35, 89; sales of, 45, 152, 165–6
Comtat Venaissin, 111, 133; institutions of, 112
conseil, see king's council
Convention, 52, 53
courts, 137; *see* king's council; judicial system; *parlements*
Craponne, Adam de, 111, 117
Craponne canal, 108
credit and irrigation, 109–10
Crillon, duke of, 116–17
Crillon canal, 104, 116–17

David, Paul, 33
Davis, Lance, 22
Desert, Gabriel, 182

213

Index

Merindol, 115–16
Merton, statute of, 164
Midlands (England), 15
Ministry of the Interior, 39, 95, 96
model of durable goods, 127–31; and grants, 140–2; and irrigation, 142–5
model of litigation, 155–63; applied to England, 163–5; applied to France, 165–8
model of water control supply, 60–2
Mont St. Michel, 41
Montpellier, 65
Morange, count of, 140
Moreille, Abbey of, 42
Mowbray, Sir John, 164–5

Nantes, 41
Napoleon, 19, 53
National Assembly, 179
Nieul, Abbey of, 42
nobles, see seigniors
Normandy, 30, 40–1, 63–5, 71, 73, 85, 92, 97–9, 181–5; drainage in, 150; failure of reform in, 88; litigation in, 139–40; royal lands in, 86
North, Douglass, 22–3, 126

Old Regime: economic performance of, 10–11; failure of, 94; institutional problems in, 36
Olivier, [?], 74–5
Olson, Mancur, 26
open fields, 14–15
Oppede family, 133, 137–9
Orgon, 117
Orléans, 16
Oursin, [?], 78, 90

Paris, 63
parlements, 30–1, 87, 117, 125, 137, 177; of Aix-en-Provence, 113, 116, 119, 134; against improvements, 47–8; litigation before, 134–5; see also judicial system
Parliament, English, 15–16
path dependence, 33–5
Pays d'Auge, 40
Pelzman, Sam, 26
Perrot, Giles, 182
Petit-Poitou marsh, 42
Physiocrats, 16
Plan-Oriental canal, 104, 120
Poitier, bishop of, 42
Poitou, 42–3
political centralization, 119
political economy, 26–8
political divisions, in Provence, 113, 119
politics, 29–30, 87

pond law, 52–3
Ponts et Chaussées, 54, 94–5, 98
Pope, 112; property of, 111
population growth, 59–60
Postel-Vinay, Gilles, 11
Postlewaite, Andrew, 152
prefects, 95–8, 119, 146
privileges, 18–19, 28, 42, 87, 111, 118, 137, 177, 179; abolition of, 96, 146; holders of, 29 (see also seigniors); as obstacles to reform, 167; see also property rights
property, peasant, 169
property rights: after 1815, 54, 172–3; to marshes, 25–6, 43–4, 72–3, 139–40, 151–3; medieval, 21, 52, 132; Old Regime, 47, 86, 174–5; reform of, 14–19, 32–3, 33–4, 47, 52, 96, 147, 166–7, 176–9; royal, 90–1; uncertain, 34–5, 46–7, 84–6, 91–4, 96–7, 153–4; to water, 138–9; see also institutions
Provence, 40, 44, 57, 63, 65, 102–3, 105, 109, 185–8; boundaries of, 100n, 113, 119; climate of, 100–1; Comté of, 111–12, 113; counts of, 111, 138; water rights in, 132
provincial authorities, see estates, provincial; intendants; parlements; prefets

Ranville, 90–1
redistribution, 15–17, 23, 176–7
reform: obstacles to, 28, 176–7; Old Regime, 17, 88; resistance to, 18, 29–32; revolutionary, 18, 95–6; see also property rights
Reinganum, Jennifer, 155
rent seeking, 23, 118
Restoration period, 97
Revolution of 1789: causes of, 9–11, 19–20, 178–9; economic effects of, 11–12, 13–14, 142, 146–8, 168, 171–2; historiography of, 1; need for, 35–6; reforms of, 18, 32–3, 95–6
Riom, 43
Rhone River, 101; delta, 65, 66
Rouen, 87

St. John of Malta, order of, 42
St. Julien, canal of, 102n, 108, 138
St. Ouen de Bures, 89
St. Radegonde, priory of, 42
St. Samson, 89
Salon, 117
Sée, Henri, 40
seigniors: disputes of, 43; against drainage, 30, 51; rights of, 94; and water control, 40–1

215